ENVIRONMENTAL COURT CASES RELATED TO BUILDINGS

DR. KAIMAN LEE
architect, landscape architect, environmental planner, real estate broker/investor/developer, personal financial/career planning consultant, researcher/writer/publisher, teacher/professor.

and

LAUREN L. KOUMJIAN
environmental researcher, ecologist.

142 Lowell Ave., Newtonville, MA. 02160

ISBN 0-915250-28-4

Printed in the United States of America

TABLE OF CONTENTS

INTRODUCTION 11

ENVIRONMENTAL COURT CASES

BASYAP VS. DISTRICT OF COLUMBIA CITY COUNCIL, NO. 99-72 19
BOSTON WATERFRONT RESIDENTS ASSOCIATION VS. ROMNEY, NO. 72-1157-LC 20
BROOKLYN RESIDENCES DEMOLITION 21
CAROLINA ACTION VS. SIMON, NO. 75-1253 22
CEDAR-RIVERSIDE VS. HILLS, NO. 4-73-592 23
CHELSEA NEIGHBORHOOD ASSOCIATIONS VS. U.S. POSTAL SERVICE, NO. 75-6005 24
CHICK VS. HILLS, NO. 75-1274 25
CITIZENS' ASSOCIATION OF GEORGETOWN VS. WASHINGTON, NO. 1944-73 26
CITY OF BOSTON VS. MASS. PORT OF AUTHORITY, NO. 7844 27
CITY OF DALLAS VS. CROWNRICH, NO. 506 S.W. 2D 654 28
CITY OF ROCHESTER VS. U.S. POSTAL SERVICE, NO. 76-6065 29
CLINTON COMMUNITY HOSPITAL CORPORATION VS. SOUTHERN MARYLAND MEDICAL CENTER, NO. 73-981-HM 30
COALITION FOR THE ENVIRONMENT VS. VOLPE, NOS. 72-1769, 72-1770 31
COMMITTEE TO SAVE THE SOUTH GREEN VS. HILLS, NO. H-76-328 32
CONCERNED ABOUT TRIDENT VS. RUMSFELD, NOS. 75-1515 AND 75-2053 33
CROSLEY BUILDING CORPORATION VS. SAMPSON, NO. 75-1043 34
CUMMINGTON PRESERVATION COMMITTEE VS. FAA, NO. 75-1198 35
DI VOSTA RENTALS, INC. VS. LEE, NO. 73-1893 36
DON'T TEAR IT DOWN, INC. VS. GSA, CIV. A, NO. 74-381 37
EAST 63RD STREET ASSOCIATION VS. COLEMAN, NO. 76, CIV. 2071 38
EDWARDS VS. FIRST BANK OF DUNDEE, NO. 75-1655 39
EDWARDS VS. FIRST BANK OF DUNDEE, NO. 393 F. SUPP. 680 40
ELY VS. VELDE, NO. 73-1889 41
FIGARSKY VS. HISTORIC DISTRICT COMMISSION, NO. 7894 42
FORT STORY - ITS FUTURE VS. SCHLESINGER, NO. 74-232-N 43
FRIENDSHIP HEIGHTS CITIZENS COORDINATING COMMITTEE VS. ROZANSKY & KAY CONSTRUCTION COMPANY 44
GETTYSBURG TOWER 45
GROTON, CONNECTICUT VS. NAVY 46
HIRAM CLARKE CIVIC CLUB, INC. VS. LYNN, NO. 72-1268 47
IRISH VS. GREEN, NO. 162-3 48
JAIL CONSTRUCTION 49
JAMES VS. LYNN, NO. 74-M-309 50
JONES VS. LYNN, NO. 73-1057 51
KENT COUNTY COUNCIL FOR HISTORIC PRESERVATION VS. ROMNEY, NO. 304 F. SUPP. 885 52
LEAGUE TO SAVE LAKE TAHOE VS. TROUNDAY 53
LITTLE WOLF LAKE PROPERTY OWNERS ASSOCIATION VS. HAASE, NO. 74-837-CE 54
MAHER VS. CITY OF NEW ORLEANS, NO. 516 F. 2D 1051 55
MARYLAND PLANNING COMMISSION VS. U.S. POSTAL SERVICE 56
MCCORMICK VS. LAWRENCE 57
MCLEAN GARDENS RESIDENTS ASSOCIATION VS. NATIONAL CAPITAL PLANNING COMMISSION 58
MILO COMMUNITY HOSPITAL VS. WEINBERGER, NO. 75-1205 59
MONTANA WILDERNESS ASSOCIATION VS. MONTANA BOARD OF HEALTH AND ENVIRONMENTAL SCIENCES, NO. 38902 60
MUHA VS. UNION LAKES ASSOCIATION, NO. 2964 61
NUCLEUS OF CHICAGO HOMEOWNERS VS. LYNN, NO. 74-1206 62

PENN CENTRAL TRANSPORTATION CO. VS. CITY OF NEW YORK, NO. 1305 63
PEOPLE OF SAIPAN VS. DEPARTMENT OF INTERIOR, NO. 73-1769 64
PINKNEY VS. OHIO ENVIRONMENTAL PROTECTION AGENCY, NO. C-73-1159 65
PLAN FOR ARCADIA VS. ANITA ASSOCIATES, NO. 74-1180 66
PLAN JEFF CO. VS. KLEPPE, NO. 76-405 67
PRINCE GEORGE'S COUNTY VS. HOLLOWAY, NO. 75-1823 68
SAN FRANCISCO TOMORROW VS. ROMNEY, C-72-65 RHS 69
SAVE OUR TEN ACRES VS. GENERAL SERVICES ADMINISTRATION 70
SAVE THE COURTHOUSE COMMITTEE VS. LYNN, NO. 74-CIV.5646 71
SCENIC RIVERS ASSOCIATION VS. LYNN, NO. 74-131-C 72
SIBSON VS. STATE, EQUITY NO. 4815 73
SIERRA CLUB VS. LYNN, NO. 73-3378 74
STATE VS. ASHMORE, NOS. 30176, 30177 75
STATE HIGHWAY COMMISSION VS. BAUMAN, NO. 28831 76
STEWART VS. DISTRICT OF COLUMBIA BOARD OF ZONING ADJUSTMENT, NO. 6802 77
SWEETWATER VALLEY CIVIC ASSOCIATION VS. CITY OF NATIONAL CITY, NO. 14408 78
3HREE LAKES ASSOCIATION VS. FISHER, NO. 1142 79
TIERRASANTA COMMUNITY COUNCIL VS. RICHARDSON, NO. 73-277-E 80
TRINITY EPISCOPAL SCHOOL VS. ROMNEY, NOS. 75-7061, 7092 81
ULSTER COUNTY COMMUNITY ACTION COMMITTEE VS. KOENIG, NO. 75-CIV.-3832 82
UNITED STATES VS. STOECO HOMES, INC., NO. 1335-72 83
WALSH VS. SPADACCIA, NO. 07182-72 84
WAMP VS. CHATTANOOGA HOUSING AUTHORITY, NO. 75-1192 85
WEST AUGUSTA HISTORICAL AND GENEALOGICAL SOCIETY VS. URBAN RENEWAL AUTHORITY OF PARKERSBURG, INC., NO. 73-2062 86
WILSON VS. LYNN, NO. 74-392-S 87

FEDERAL ACTS

1. The National Environmental Policy Act of 1969, as amended 89
2. Council on Environmental Quality: Preparation of Environmental Impact Statements - Guidelines 94

INDEX — BUILDING TYPES

APARTMENT BUILDING

CITY OF DALLAS VS. CROWNRICH 28
DI VOSTA RENTALS, INC. VS. LEE 36
HIRAM CLARKE CIVIC CLUB, INC. VS. LYNN 47
MAHER VS. CITY OF NEW ORLEANS 55
MCLEAN GARDENS RESIDENTS ASSOCIATION VS. NATIONAL CAPITAL PLANNING COMMISSION 58
WALSH VS. SPADACCIA 84
WAMP VS. CHATTANOOGA HOUSING AUTHORITY 85

COMMERCIAL BUILDING COMPLEX (RESTAURANT, HOTEL, SHOPPING CENTER)

CITIZENS ASSOCIATION OF GEORGETOWN VS. WASHINGTON 26
MCLEAN GARDENS RESIDENTS ASSOCIATION VS. NATIONAL CAPITAL PLANNING COMMISSION 58
MUHA VS. UNION LAKES ASSOCIATION 61

CONDOMINIUM

FRIENDSHIP HTS. CITIZENS COORDINATING COMMITTEE VS. ROZANSKY AND KAY CONSTRUCTION COMPANY 44
STATE VS. ASHMORE 75
STATE HIGHWAY COMMISSION VS. BAUMAN 76

DETENTION CENTER

JAIL CONSTRUCTION 49
TIERRASANTO COMMUNITY COUNCIL VS. RICHARDSON 80

GOVERNMENT BUILDING (JUSTICE BUILDING AND CITY HALL)

CAROLINA ACTION VS. SIMON 22

HOSPITAL

CLINTON COMMUNITY HOSPITAL CORP. VS. SO. MARYLAND MEDICAL CENTER 30
ELY VS. VELDE 41
MILO COMMUNITY HOSPITAL VS. WEINBURGER 59

HOTEL

LEAGUE TO SAVE LAKE TAHOE VS. TROUNDAY 53
PEOPLE OF SAIPAN VS. DEPARTMENT OF INTERIOR 64

MEDICAL CENTER

ELY VS. VELDE 41

OBSERVATION TOWER

GETTYSBURG TOWER 45

OFFICE BUILDING (HIGHRISE)

BASYAP VS. DISTRICT OF COLUMBIA 19
CROSLEY BUILDING CORPORATION VS. SAMPSON 34
PENN CENTRAL TRANSPORTATION CO. VS. CITY OF NEW YORK 63
SAVE OUR TEN ACRES VS. GENERAL SERVICES ADMINISTRATION 70

OFFICE BUILDING (LOWRISE)

EDWARDS VS. FIRST BANK OF DUNDEE 39,40
PLAN JEFF CO VS. KLEPPE 67

PARKING FACILITY

CITY OF BOSTON VS. MASS. PORT AUTHORITY 27
FRIENDSHIP HEIGHTS CITIZENS COORDINATING COMMITTEE VS. ROZANSKY AND KAY CONSTRUCTION COMPANY 44
MUHA VS. UNION LAKES ASSOCIATION 61

PLANNED COMMUNITY

COALITION FOR THE ENVIRONMENT VS. VOLPE 31
ULSTER COUNTY COMMUNITY ACTION COMMITTEE VS. KOENIG 82

PLANT DEVELOPMENT

BROOKLYN RESIDENTS DEMOLITION 21

POSTAL FACILITY

CITY OF ROCHESTER VS. U.S. POSTAL SERVICE 29
MARYLAND PLANNING COMMISSION VS. U.S. POSTAL SERVICE 56

RADAR FACILITY

CUMMINGTON PRESERVATION COMMITTEE VS. FAA 35

RECREATIONAL FACILITY

STEWART VS. DISTRICT OF COLUMBIA BOARD OF ZONING ADJUSTMENT 77

RESEARCH CENTER

PRINCE GEORGE'S COUNTY VS. HOLLOWAY 68

RESIDENTIAL HOUSING PROJECT (MULTIPLE DWELLINGS)

CEDAR-RIVERSIDE VS. HILLS 23
FORT STORY - ITS FUTURE VS. SCHLESINGER 43
GROTON, CONNECTICUT VS. NAVY 46
MONTANA WILDERNESS ASSOCIATION VS. MONTANA BOARD OF HEALTH AND ENVIRONMENTAL SCIENCES 60
SIERRA CLUB VS. LYNN 74
TRINITY EPISCOPAL SCHOOL VS. ROMNEY 81

RESIDENTIAL HOUSING (SINGLE HOMES)

CHICK VS. HILLS 25
FIGARSKY VS. HISTORIC DISTRICT COMMISSION 42
SIBSON VS. STATE 73

SHOPPING CENTER

MUHA VS. UNION LAKES ASSOCIATION 61
PINKNEY VS. OHIO ENVIRONMENTAL PROTECTION ACENCY 65
PLAN FOR ARCADIA VS. ANITA ASSOCIATES 66
SWEETWATER VALLEY CIVIC ASSOCIATION VS. CITY OF NATIONAL CITY 78

SUBWAY STATION

EAST 63RD STREET ASSOCIATION VS. COLEMAN 38

URBAN RENEWAL PROJECT

COMMITTEE TO SAVE THE SOUTH GREEN VS. HILLS 32
DON'T TEAR IT DOWN, INC. VS. GSA 37
JAMES VS. LYNN 50
JONES VS. LYNN 51
KENT COUNTY COUNCIL FOR HISTORIC PRESERVATION VS. ROMNEY 52
SAN FRANCISCO TOMORROW VS. ROMNEY 69
SAVE THE COURTHOUSE COMMITTEE VS. LYNN 71
WEST AUGUSTA HISTORICAL AND GENEALOGICAL SOCIETY VS. URBAN RENEWAL 86
WILSON VS. LYNN 87

VEHICLE FACILITY HOUSING PROJECT - PARKING GARAGE

CHELSEA NEIGHBORHOOD ASSOCIATION VS. U.S. POSTAL SERVICE 24
CITY OF BOSTON VS. MASS. PORT AUTHORITY 27
FRIENDSHIP HEIGHTS CITIZENS COORDINATING COMMITTEE VS. ROZANSKY AND KAY CONSTRUCTION COMPANY 44

WATERFRONT RESIDENTIAL DEVELOPMENT

BOSTON WATERFRONT RESIDENTS ASSOCIATION VS. ROMNEY 20
IRISH VS. GREEN 48
LITTLE WOLF LAKE PROPERTY OWNERS ASSOCIATION VS. HAASE 54
MCCORMICK VS. LAWRENCE 57
MUHA VS. UNION LAKES ASSOCIATION 61
SCENIC RIVERS ASSOCIATION VS. LYNN 72
SIBSON VS. STATE 73
STATE HIGHWAY COMMISSION VS. BAUMAN 76
THREE LAKES ASSOCIATION VS. FISHER 79
UNITED STATES VS. STOECO HOMES, INC. 83

INDEX — ENVIRONMENTAL ISSUES

AESTHETICS

CLINTON COMMUNITY HOSPITAL CORPORATION VS. SOUTHERN MARYLAND MEDICAL CENTER 30
IRISH VS. GREEN 48
MCCORMICK VS. LAWRENCE 57
MONTANA WILDERNESS ASSOCIATION VS. MONTANA BOARD OF HEALTH AND ENVIRONMENTAL SCIENCES 60

AIR QUALITY STANDARDS

CITIZEN'S ASSOCIATION OF GEORGETOWN VS. WASHINGTON 26
CITY OF BOSTON VS. MASS. PORT. AUTHORITY 27
COALITION FOR THE ENVIRONMENT VS. VOLPE 31
FRIENDSHIP HEIGHTS CITIZENS COORDINATING COMMITTEE VS. ROZANSKY AND KAY CONSTRUCTION COMPANY 44
LEAGUE TO SAVE LAKE TAHOE VS. TROUNDAY 53
MCLEAN GARDENS RESIDENTS ASSOCIATION VS. NATIONAL CAPITAL PLANNING 58
PINKNEY VS. OHIO ENVIRONMENTAL PROTECTION AGENCY 65
PLAN FOR AREADIA VS. ANITA ASSOCIATES 66
PLAN JEFF CO VS. KLEPPE 67
SAVE OUR TEN ACRES VS. GENERAL SERVICES ADMINISTRATION 70

ALTERNATIVES

CUMMINGTON PRESERVATION COMMITTEE VS. FAA 35
MONTANA WILDERNESS ASSOCIATION VS. MONTANA BOARD OF HEALTH AND ENVIRONMENTAL SCIENCES 60
NUCLEUS OF CHICAGO HOMEOWNERS VS. LYNN 62
SWEETWATER VALLEY CIVIC ASSOCIATION VS. CITY OF NATIONAL CITY 78
TRINITY EPISCOPAL SCHOOL VS. ROMNEY 81

BUILDING PERMITS

CITY OF DALLAS VS. CROWNRICH 28
PINKNEY VS. OHIO ENVIRONMENTAL PROTECTION AGENCY 65
UNITED STATES VS. STOECO HOMES, INC. 83

CRIME

EAST 63RD STREET ASSOCIATION VS. COLEMAN 38
NUCLEUS OF CHICAGO HOMEOWNERS VS. LYNN 62

DEMOLITION OF BUILDINGS

BOSTON WATERFRONT RESIDENTS ASSOCIATION VS. ROMNEY 20
BROOKLYN RESIDENCES DEMOLITION 21
COMMITTEE TO SAVE THE SOUTH GREEN VS. HILLS 32
DON'T TEAR IT DOWN, INC. VS. GSA 36
EDWARDS VS. FIRST BANK OF DUNDEE 39,40
FIGARSKY VS. HISTORIC DISTRICT COMMISSION 42
JAMES VS. LYNN 50
JONES VS. LYNN 51
MCLEAN GARDENS RESIDENTS ASSOCIATION VS. NATIONAL CAPITAL PLANNING 58
WEST AUGUSTA HISTORICAL AND GENEALOGICAL SOCIETY VS. URBAN RENEWAL AUTHORITY OF PARKERSBURG, INC. 86

ECONOMIC CONSIDERATIONS

CEDAR-RIVERSIDE VS. HILLS 23
EAST 63RD STREET ASSOCIATION VS. COLEMAN 38
MONTANA WILDERNESS ASSOCIATION VS. MONTANA BOARD OF HEALTH AND ENVIRONMENTAL SCIENCES 60
NUCLEUS OF CHICAGO HOMEOWNERS VS. LYNN 62
PLAN JEFF CO VS. KLEPPE 67

EIS ADEQUACY

CEDAR-RIVERSIDE VS. HILLS 23
CITY OF BOSTON VS. MASS. PORT AUTHORITY 27
CUMMINGTON PRESERVATION COMMITTEE VS. FAA 35
EAST 63RD STREET ASSOCIATION VS. COLEMAN 38
JAMES VS. LYNN 50
MONTANA WILDERNESS ASSOCIATION VS. MONTANA BOARD OF HEALTH AND ENVIRONMENTAL SCIENCES 60
PRINCE GEORGE'S COUNTY VS. HOLLOWAY 68
SIERRA CLUB VS. LYNN 74
SWEETWATER VALLEY CIVIC ASSOCIATION VS. CITY OF NATIONAL CITY 78

ENVIRONMENTAL RIGHTS

GETTYSBURG TOWER 45
PINKNEY VS. OHIO ENVIRONMENTAL PROTECTION AGENCY 65

FUTURE CONSIDERATIONS

MONTANA WILDERNESS ASSOCIATION VS. MONTANA BOARD OF HEALTH AND ENVIRONMENTAL SCIENCES 60

GRANT FUNDING

CAROLINA ACTION VS. SIMON 22
ELY VS. VELDE 41
ULSTER COUNTY COMMUNITY ACTION COMMITTEE VS. KOENIG 82

HIGH DENSITY DEVELOPMENT

CEDAR-RIVERSIDE VS. HILLS 23
COALITION FOR THE ENVIRONMENT VS. VOLPE 31
FORT STORY - ITS FUTURE? VS. SCHLESINGER 43
PRINCE GEORGE'S COUNTY VS. HOLLOWAY 68
THREE LAKES ASSOCIATION VS. FISHER 79

HISTORIC (NHPA) SIGNIFICANCE

BOSTON WATERFRONT RESIDENTS ASSOCIATION VS. ROMNEY 20
COMMITTEE TO SAVE THE SOUTH GREEN VS. HILLS 32
DON'T TEAR IT DOWN VS. GSA 37
EDWARDS VS. FIRST BANK OF DUNDEE 39,40
ELY VS. VELDE 41
FIGARSKY VS. HISTORIC DISTRICT COMMISSION 42
KENT COUNTY COUNCIL FOR HISTORIC PRESERVATION VS. ROMNEY 52
MAHER VS. CITY OF NEW ORLEANS 55
SAVE THE COURTHOUSE COMMITTEE VS. LYNN 71
WEST AUGUSTA HISTORICAL AND GENEALOGICAL SOCIETY VS. URBAN RENEWAL AUTHORITY OF PARKERSBURG, INC. 86

LAND FILL PERMITS

DI VOSTA RENTALS INC. VS. LEE 36
SIBSON VS. STATE 73

LAND LEASING

CROSLEY BUILDING CORPORATION VS. SAMPSON 34
MAHER VS. CITY OF NEW ORLEANS 55
PLAN JEFF CO VS. KLEPPE 67

LAND USE PLANNING

COALITION FOR THE ENVIRONMENT VS. VOLPE 31
GETTYSBURG TOWER 45
IRISH VS. GREEN 48
MARYLAND PLANNING COMMISSION VS. U.S. POSTAL SERVICE 56

NEPA APPLICABILITY

BASYAP VS. DISTRICT OF COLUMBIA CITY COUNCIL 19
BOSTON WATERFRONT RESIDENTS ASSOCIATION VS. ROMNEY 20
BROOKLYN RESIDENTS DEMOLITION 21
CAROLINA ACTION VS. SIMON 22
CHELSEA NEIGHBORHOOD ASSOCIATIONS VS. U.S. POSTAL SERVICE 24
CITY OF ROCHESTER VS. U.S. POSTAL SERVICE 29
COMMITTEE TO SAVE THE SOUTH GREEN VS. HILLS 32
CROSLEY BUILDING CORPORATION VS. SAMPSON 34
ELY VS. VELDE 41
FORT STORY - ITS FUTURE VS. SCHLESINGER 43
GROTON, CONNECTICUT VS. NAVY 46
HIRAM CLARKE CIVIC CLUB VS. LYNN 47
IRISH VS. GREEN 48
JAIL CONSTRUCTION 49
JONES VS. LYNN 51
MARYLAND PLANNING COMMISSION VS. U.S. POSTAL SERVICE 56
MCLEAN GARDENS RESIDENTS ASSOCIATION VS. NATIONAL CAPITAL PLANNING 58
MILO COMMUNITY HOSPITAL VS. WEINBERGER 59
NUCLEUS OF CHICAGO HOMEOWNERS VS. LYNN 62
PEOPLE OF SAIPAN VS. DEPARTMENT OF INTERIOR 64
PLAN JEFF CO VS. KLEPPE 67
SAN FRANCISCO TOMORROW VS. ROMNEY 69
SAVE OUR TEN ACRES VS. GENERAL SERVICES ADMINISTRATION 70
SAVE THE COURTHOUSE COMMITTEE VS. LYNN 71
SCENIC RIVERS ASSOCIATION VS. LYNN 72
TIERRASANTO COMMUNITY COUNCIL VS. RICHARDSON 80
TRINITY EPISCOPAL SCHOOL VS. ROMNEY 81
ULSTER COUNTY COMMUNITY ACTION COMMITTEE VS. KOENIG 82
WAMP VS. CHATTANOOGA HOUSING AUTHORITY 85
WILSON VS. LYNN 87

NOISE

COALITION FOR THE ENVIRONMENT VS. VOLPE 31
PLAN JEFF CO VS. KLEPPE 67

PLAN ALTERATIONS AFTER EIS APPROVAL

CHICK VS. HILLS 25
WILSON VS. LYNN 87

PRIVATE PROPERTY RIGHTS

MAHER VS. CITY OF NEW ORLEANS 55
PENN CENTRAL TRANSPORTATION CO. VS. CITY OF NEW YORK 63
SIBSON VS. STATE 73
STATE VS. ASHMORE 75
STATE HIGHWAY COMMISSION VS. BAUMAN 76

REVENUE SHARING

CAROLINA ACTION VS. SIMON 22

SOCIAL IMPACT

CITY OF ROCHESTER VS. U.S. POSTAL SERVICE 29
EDWARDS VS. FIRST BANK OF DUNDEE 39,40
JAIL CONSTRUCTION 49
MCLEAN GARDENS RESIDENTS ASSOCIATIONS VS. NATIONAL CAPITAL PLANNING COMMISSION 58
NUCLEUS OF CHICAGO HOMEOWNERS VS. LYNN 62
PLAN JEFF CO VS. KLEPPE 67
TIERRASANTO COMMUNITY COUNCIL VS. RICHARDSON 80

TRAFFIC

CITY OF ROCHESTER VS. U.S. POSTAL SERVICE 29
IRISH VS. GREEN 48
MCLEAN GARDENS RESIDENTS ASSOCIATION VS. NATIONAL CAPITAL PLANNING COMMISSION 58
PINKNEY VS. OHIO ENVIRONMENTAL PROTECTION AGENCY 65
PLAN JEFF CO VS. KLEPPE 67
SAVE OUR TEN ACRES VS. GENERAL SERVICES ADMINISTRATION 70
WALSH VS. SPADACCIA 84

WATER QUALITY

IRISH VS. GREEN 48
MUHA VS. UNION LAKES ASSOCIATION 61
SCENIC RIVERS ASSOCIATION VS. LYNN 72
SIERRA CLUB VS. LYNN 74
THREE LAKES ASSOCIATION VS. FISHER 79
UNITED STATES VS. STOECO HOMES, INC. 83
WALSH VS. SPADACCIA 84

ZONING CODES

CITY OF DALLAS VS. CROWNRICH 28
LITTLE WOLF LAKE PROPERTY OWNERS ASSOCIATION VS. HAASE 54
MAHER VS. CITY OF NEW ORLEANS 55
MARYLAND PLANNING COMMISSION VS. POSTAL SERVICE 56
STEWART VS. DISTRICT OF COLUMBIA BOARD OF ZONING ADJUSTMENT 77
TIERRASANTA COMMUNITY COUNCIL VS. RICHARDSON 80
WALSH VS. SPADACCIA 84

INTRODUCTION

Since January 1, 1970, when the National Environmental Policy Act (NEPA) went into effect, the law had generated a tremendous number of litigations for the federal courts. Although only a few of the suits had resulted in cancellation of federal projects on environmental grounds, the Council on Environmental Quality (CEQ) reported that NEPA had signigicant impact on federal agency decision-making.

In 1973, over 2,000 environmental impact statements were prepared at the faderal level and submitted to the Council on Environmental Quality. Their preliminary conclusion was that the majority of the impact statements were not prepared well enough in presenting sufficient information to allow a neutral decision-maker to judge the importance of the environmental benefits and costs of the proposed projects, thus resulted in delaying the federal agancy's decision-making and costing the public's tax money. One might then conclude that either many consulting firms and agencies were poorly prepared and/or were ignoring the mandate of the National Environmental Policy Act.

Environmental impact statements when good can be very good - both in helping the developer to build an economically and environmentally sound project and providing planners and local officials with data for making well-informed decisions. When bad, the environmental impact statement can cost time and money, not only during the approval process, but later on in the project's life when unanticipated effects bring grief to consumer, public agency, and, sometimes, developer alike. Nevertheless, despite all the lists, handbooks, guidelines, and manuals available, it is almost as difficult to define a "good" environmental statement as it is to write one.

Preparation of a good assessment is no guarantee for speedy processing. Inadequate assessments are often accepted without a murmur, while exhaustive studies may well result in delays if only because they make difficult reading. Before embarking on the preparation of an impact statement, therefore, the developer, public planner, or consultant should be aware not only of the legal requirements, but also the special interests and aspirations of the statement's potential audience. Taking on these precautions will ultimately reduce the chance of fighting the case in court.

The courts' leading role in NEPA may be traced in large measure to their current willingness to review all agency action more closely than they did a few years ago. Congress failed to specify in detail how the NEPA Act should be implemented. As a result, the courts are thought of as the principal enforcers of NEPA. Moreover, NEPA implicitly confers authority on the federal courts to enjoin any federal action taken in violation of NEPA's procedural requirements, even if jurisdiction to review this action is otherwise lacking.

NEPA has been generally accorded a sympathetic reception by the courts. For several years the courts have been attempting to evolve better standards of review from existing statutory and case law. They have asked that agencies take a wider array of public interests into account in carrying out their statutory missions, that agencies consider alternatives to their proposals, that they articulate the grounds for their decisions, that they spell out procedures for principle decision-making, and that they provide for public participation of various kinds.

Thus judicial review and the NEPA process share many of the same goals. By policing NEPA, the courts are able to achieve partial review of the quality of agency decision-making by enforcing a statute which Congress enacted to bring about reforms in the administrative process. At the same time, they are protecting public interests in life, health, and the environment from administrative arbitrariness.

When NEPA first went into effect it created a great deal of confusion among the agencies about the "retroactivity" of its provisions, with reference to agency projects which had been started prior to January 1, 1970, but were not yet complete. This is particularly important in terms of the ability to require an EIS filing for major projects. Although a few of the early decisions flatly refused to apply the requirements of the act to a project begun before its effective date, within two years most courts were applying a "balancing test" to determine whether the federal law should apply to a given project.

It will usually be decided that the law is applicable to an ongoing project if there is still a substantial amount of work to be done, and if, realistically, there is an opportunity to reconsider (if necessary) aspects of that remaining work. Thus, where the benefits are not outweighed by the environmental costs, and the initiator will not suffer irreparable harm while an EIS is prepared for the ongoing project, or remaining segments of it, courts have usually issued injunctions halting all work until the EIS is prepared. Naturally, as the years pass, this will tend to become a dead issue for federal agency filings, but increasingly important for local filings under a variety of ordinance formats.

As to the question of adequacy of an EIS, there are four main areas which have been used by the courts as a basis for finding a particular EIS inadequate: lack of detail, failure to discuss alternatives, failure to consider opposing views, and obscuring the environmental effects of the proposed project.

It should go without saying that the mere filing of a document labelled "final impact statement" is insufficient to shield an agency from judicial review if the document fails to comply with the standards outlined in NEPA.

In order to avoid rejection as insufficiently detailed or overly conclusory, the statement must set forth not only the views and conclusions of the agency, but also all those brought to its attention during the commenting process, including those claims made by opponents of the undertaking. Even if the agency disagrees violently with a particular point of view, it should include it in the final statement.

The practice of measuring adequacy by the bench mark of "full disclosure" is now standard. Useful as it is, the standard of full disclosure raises the question of how full is "full." NEPA has been interpreted to require both the sharing of information with the public and the involvement of the public in developing information.

A statement's adequacy, in the end, is measured by its functional usefulness in decision making. The statement must be of a nature and form that enables the decision maker to consider environmental factors in good faith. Consideration through a "finely tuned and 'systematic' balancing process" must take place.

The Senate discussion of NEPA clearly indicates that the Act was intended to facilitate policy choices, yet there can be no choice if a decision maker lacks alternatives to the proposed action. The equivalent of presenting alternatives is that agencies develope "appropriate alternatives" when there are "unresolved conflicts concerning alternative uses of available resources." These two sections of the statute tell agencies to present reviewers with options other than the one favored by the agency. It is not surprising, then, that the agencies have been challenged on grounds of excluding relevant alternatives from statements and inadequately discussing those presented.

The alternatives of no action at all, and of another action that fully accomplishes the original goal but without any of its objectionable features, constitute the extremes between which discussable alternatives fall.

In between no action at all and the perfect alternative lies a wide range of more likely choices that offer partial solutions or go only partway toward meeting the original proposal's objectives. As important as the range of alternatives is the thoroughness with which each alternative must be discussed.

Since the function of the EIS is to alert agency decision-makers to the environmental consequences of a proposed project, the final EIS has to include responsible opposing views, and must in turn answer those comments. This is much more a public rather than private responsibility regardless of who initiates the EIS. Once the necessity for such information has been made clear to the agencies (by the couts), the agencies have generally responded with alacrity.

Because the size of the usual EIS is enormous, at a minimum a thick single volume, it would be relatively easy to hide the environmental effects of a project in a mass of verbiage, to bury the impacts in a welter of detail. One court even had to go so far as to establish guidelines for the preparation of the EIS, so that the court, when reviewing the EIS could clearly understand the impact of each segment of the decades-long project, and its relation to the whole, as well as the eventual total impact. In fact, brutal frankness about the environmental effects may at times be an asset. The corps of Engineers bluntly and candidly summarized the effect of one of its projects in this way:

> "Practically no natural, productive wildlife habitat will survive if the project is implemented, and there will be a permanent loss to present recreational uses, and aesthetic qualities afforded by the open natural area."

The EIS was found to be adequate.

Once the EIS has been found to be adequate, those opposing the project have only one recourse left: to seek review of the agency decision on substantive, rather than procedual, grounds. Until November 1972, every decision construing NEPA had concluded that it was a procedural law only, and that once an agency had complied with the procedural requirements, the courts' function was at an end.

No one really needs to be convinced that the requirement for impact statements has dominated NEPA's interpretation by the courts. Almost every word of 102(2) (C) has been construed several times, with important variations in shades of meaning which have sometimes produced inconsistent results.

The courts have been active in reviewing agency compliance with NEPA, in large part because private citizens interested in NEPA's enforcement have been granted "standing" to sue. As "proper parties to request adjudication of a particular issue," citizens' groups have provided the courts with numerous opportunities to rule on NEPA's implementation. Few challenges to citizens' standing have succeeded.

Standing under NEPA is the same as standing under any other legislation; however, a special aspect of standing under NEPA deserves special mention. NEPA's enactment arguably creates new legal interests which enlarge the category of injuries which the public may sustain, so that harm to the public's right to know, to participate, and to have the interests of future generations protected may constitute injury in fact.

Economic injury to property owners from proposed projects, of course, is sufficient to confer standing.

Another category of cognizable injury consisting of injury to "aesthetic, conservational, or recreational interests," has given the courts somewhat more trouble.

What have the few years of litigation under NEPA accomplished? The cases that we will have discussed have not accomplished very much in achieving NEPA's ultimate goal of a better environment through better federal decision making. However, due to the courts and the strict enforcement of NEPA's procedural requirements, they have gone a long way toward ensuring that adequate agency decision-making machinery exists so that better substantive results may be obtained in the future, and in some instances, have led to useful debate and discussion about them.

Through review of the adequacy of impact statements, the courts exercised a large measure of control over the thoroughness and competency of the environmental assessments made. If the foregoing accomplishments did not ensure different agency decisions, they at least established that facts which might support different decisions had to be disclosed and discussed.

Let us now look at the future trends. There is a perceptible trend toward litigation involving the subtleties and nuances of NEPA. For instance, take the key NEPA phrase: "a major federal action significantly affecting the quality of the human environment." There has been a tremendous amount of litigation directly or indirectly involving the words "major" and "significant." Only now, a dew cases are beginning to appear which take a closer look at the phrase, "human environment." A majority of NEPA cases concerns man's physical environment: losses to birds and bees, to flowers and trees, to wildlife and open spaces. There is almost an aura of warfare between endangered living things, and technology in the name of "progress."

The remaining cases are evidencing a concern for man's social and psychological environment. For example, a case that conseders the effect of a jail across the street from an apartment area upon the "human environment" of the neighborhood, another case that considers the alleged negative sociological and psychological impact of tenants of public housing on a middle-class neighborhood, or a case that considers the impact of a 206-acre youth correctional facility within 500 feet of a proposed elementary school and essentially adjoining, or partially surrounded by, a low-density single-family residential neighborhood.

Another facet beginning to surface is concern for cumulative environmental impacts. The court ordered the agencies to prepare not only an EIS for each identifiable segment of the project, but for the project as a whole, showing how each segment relates to the totality, and how the environmental impact of each segment relates to the total impact when the project is completed. A community will take many years to build, and it is entirely possible that an environmental analysis of each identifiable segment as it was constructed might show a balance of environmental benefits over costs, but unless the entire project is evaluated at the start, the cumlative negative impacts (if any) may build up so that, without such early evaluation, by the time the project is completed and the negative impacts are perceived nothing could be done to rectify them.

This book encompasses precedental court decisions conserning the environment and buildings since the enactment of the NEPA. Each case included in the text includes a summary of the issue(s) and the court decision as well as separate headings for Building Type, Location, Issue(s), and Reference. Three indices direct the reader to specific cases by title of the case,building type, and environmental issue(s).

Specific details of the case, naturally, are not included in each summary. Instead, the user can get synopses of many cases without the tedious task of briefing the entire court proceedings. Also, the reader benefits by having all the significant court cases involving federal building projects and the environment form the inception of the NEPA through 1976 in one reference book. In this way, the text is a convenient and concise reference for the lawyer concerned with achieving a balance between construction within the environment and protection of that environment. For the student of law the text follows the development of federal court prodedure and decision making for a law which does not deal specifically with violations. Finally, the text can be useful to the planner/builder in providing an insight to specific issues that bring a federal construction agency to the courts. The architect that accepts the terms of the NEPA and understands how they work in specific instances will follow a smoother path than the architect that does not seriously consider the consequences. On the one hand, if this text is referred to <u>before</u> planning a construction, the planner can avoid lengthy, expensive court proceedings. On the other hand, should the architect find himself facing court charges, the reference will enable him to talk intelligently about the issue(s).

It is hoped that anyone reading this text will appreciate its conciseness, clarity, and, above all, its importance to an environmentally aware society that all of us are advancing.

ENVIRONMENTAL COURT CASES

CASE	BASYAP VS. DISTRICT OF COLUMBIA CITY COUNCIL, NO. 99-72
LOCATION	WASHINGTON, D.C.
DATE & STATUS	ENJOINED APRIL 26, 1972
PROJECT TYPE	HIGHRISE OFFICE BUILDINGS, PART OF URBAN RENEWAL PROJECT
ISSUE(S)	NEPA APPLICABILITY, AESTHETICS, TRAFFIC, AIR POLLUTION
SCOPE	PLAINTIFFS SOUGHT TO COMPEL DEFENDANTS (DISTRICT OF COLUMBIA REDEVELOPMENT LAND AGENCY) TO PREPARE A NEPA IMPACT STATEMENT BEFORE ACQUIRING PROPERTIES IN URBAN RENEWAL AREAS. PLAINTIFF ARGUED THAT DEMOLITION OF EXISTING STRUCTURES AND THEIR REPLACEMENT WITH HIGHRISE OFFICE BUILDINGS WOULD HAVE ADVERSE AESTHETIC EFFECTS, OBSCURE LIGHT FROM THE STREET LEVEL, TRAP AND POLLUTED AIR, AND BRING NEW MASSES OF PEOPLE AND VEHICLE TRAFFIC TO THE AREA. THE COURT FOUND THE DOWNTOWN URBAN RENEWAL PROJECT A MAJOR FEDERAL ACTION SIGNIFICANTLY AFFECTING THE QUALITY OF THE HUMAN ENVIRONMENT, AND SO REQUIRING THE CONSIDERATION OF ENVIRONMENTAL IMPACTS AT EVERY IMPORTANT STAGE IN THE DECISION MAKING PROCESS.
REFERENCE	ENVIRONMENTAL LAW REPORTER, DIG. VOL. 2, P 65213

CASE	BOSTON WATERFRONT RESIDENTS ASSOCIATION VS. ROMNEY, NO. 72-1157-LC
LOCATION	BOSTON, MASSACHUSETTS
DATE & STATUS	ENJOINED ON APRIL 18, 1972
PROJECT TYPE	WATERFRONT RESIDENTIAL REDEVELOPMENT
ISSUE(S)	DESTRUCTION OF BUILDINGS WITH HISTORICAL SIGNIFICANCE, NEPA APPLICABILITY
SCOPE	THE COURT RULED THAT THE BOSTON WATERFRONT REDEVELOPMENT PROJECT IS A MAJOR ACTION AND REQUIRED AN IMPACT STATEMENT UNDER NEPA. RESTRAINING THE BOSTON REDEVELOPMENT AUTHORITY FROM PROCEEDING WITH THE DEMOLITION OF THE BUILDINGS. THE PLAINTIFFS CHARGED THAT HUD AND THE BRA FAILED TO COMPLY WITH REGULATIONS PROMULGATED BY HUD FOR CONSULTING RESIDENTS IN AREAS SUBJECT TO RESIDENTIAL REHABILITATION AND FAILED TO FILE AN ENVIRONMENTAL IMPACT STATEMENT AS REQUIRED BY NEPA. THERE WAS CONTROVERSY OVER THE HISTORICAL VALUE OF THE BUILDINGS, WHICH HAVE BEEN PROPOSED FOR LISTING ON THE NATIONAL REGISTER OF HISTORIC LANDMARKS. THE BRA HAD ORDERED THEIR DEMOLITION BECAUSE OF THEIR CONDITION, BUT IT WAS NOTED THAT FOR THE PURPOSES OF SAFETY OTHER METHODS COULD BE TEMPORARILY EMPLOYED, WHILE THE DANGEROUS CONDITION COULD BE REMEDIED BY RESTORATION OF THE BUILDINGS. THE DEFENDENTS ARGUED THAT SINCE THE PROJECT WAS APPROVED AND THE FUNDS ALLOCATED PRIOR TO THE PASSAGE OF NEPA, ANY ACTIONS AS A PART OF THIS PROJECT ARE NOT GOVERNED BY THAT STATUTE. IN ITS OPINION THE COURT SAID, "THE WATERFRONT PROJECT IS A MAJOR ONE WHICH WILL EXTEND INTO THE FUTURE AND WHICH INVOLVES CONTINUING FEDERAL INVOLVEMENT ON A LARGE SCALE. IN LIGHT OF THE STRONG POLICY DECLARATIONS OF NEPA THE COURT CANNOT FIND THAT SUCH A PROJECT SHOULD BE EXEMPTED FROM THE STATUTE, WHEN PARTICULAR PARTS OF THE PROJECT ARE STILL PLANNED FOR THE FUTURE AND WHEN THERE IS STILL TIME AND OPPORTUNITY FOR THE CONSIDERATION OF ALTERNATIVES." WHILE NOT DECIDING ON THE HISTORIC VALUE OF THE BUILDINGS, THE COURT NOTED THE IRREVOCABILITY OF THE ACT OF DEMOLITION. THE COURT THEREFORE RESTRAINED THE BRA FROM PROCEEDING WITH THE DEMOLITION UNTIL FURTHER ORDER OF THE COURT.
REFERENCE	ENVIRONMENTAL LAW REPORTER, VOL. 2, PP 20359-20360

CASE	BROOKLYN RESIDENCES DEMOLITION
LOCATION	BROOKLYN, NEW YORK
DATE & STATUS	REINSTATED ON 19 SEPTEMBER, 1973
PROJECT TYPE	PLANT EXPANSION
ISSUE(S)	LACK OF IMPACT STATEMENT, DEMOLITION OF RESIDENCES

SCOPE

THE U.S. COURT OF APPEALS FOR THE SECOND CIRCUIT REFUSED TO ISSUE, ON THE BASIS OF ALLEGED VIOLATIONS OF THE NEPA, A PRELIMINARY INJUNCTION HALTING THE DEMOLITION OF RESIDENCES IN NEW YORK CITY.

THE CASE WAS FILED BY BROOKLYN RESIDENTS WHO HAVE BEEN EVICTED OR FACE EVICTION BY THE CITY, SO THAT THE CITY CAN RAZE THE SITE UNDER ITS POWER OF CONDEMNATION. THE AREA WIL BE THE SITE OF S&S CORRUGATED PAPER MACHINERY COMPANY'S EXPANDED PLANT WHICH IS BEING PARTIALLY FINANCED BY A LOAN FROM THE COMMERCE DEPARTMENT'S ECONOMIC DEVELOPMENT ADMINISTRATION (EDA). THE PLAINTIFFS CHARGED THAT EDA FAILED TO PREPARE AN ENVIRONMENTAL IMPACT STATEMENT WHEN IT ISSUED THE LOAN COMMITMENT.
EDA FAILED TO PREPARE AN ENVIRONMENTAL IMPACT STATEMENT WHEN IT ISSUED THE LOAN COMMITMENT.

PLAINTIFFS ARGUED THAT THE CITY COULD BE ENJOINED, EVEN THOUGH IT WAS NOT THE BENEFICIARY OF THE LOAN, SINCE IT WAS "INDIRECTLY" THE BENEFICIARY OF THE LOAN, OR WAS IN A SENSE "IN PARTNERSHIP" WITH EDA.

THE COURT NOTED, HOWEVER, THAT THE CITY'S CONDEMNATION AND POSSESSION OF THE SITE FOR THE PROPOSED PLANT DOES NOT LEGALLY RELY ON EDA'S LOAN COMMITMENT AND THAT THE INITIAL PHASE OF THE CONSTRUCTION WILL IN FACT PROCEED WITHOUR EDA FUNDS, RULING AGAINST PLAINTIFFS. THE COURT DID NOT DECIDE WHETHER A NONRECIPIENT OF A LOAN MAY EVER BE IN PARTNERSHIP WITH THE FEDERAL GOVERNMENT.

JUDGE JAMES L. OAKES, WRITING FOR THE COURT, FOUND THAT EDA'S APPROVAL OF THE LOAN WAS CLEARLY A MAJOR FEDERAL ACTION, BUT THAT ISSUANCE OF A PRELIMINARY INJUNCTION WAS NOT WARRANTED, SINCE PLAINTIFFS' SHOWING OF IRREPARABLE HARM RELATED ONLY TO THEIR EVICTION BY THE CITY AND NOT TO ACTUAL PLANT CONSTRUCTION. NOTING THAT THE PROBABLE TIME FOR DISBURSEMENT OF THE FUNDS IS 1974, HE RULED THAT ANY HARM TO THE ENVIRONMENT CAUSED BY THE CONSTRUCTION WILL NOT ACCRUE UNTIL AFTER A FULL TRIAL ON THE MERITS CAN BE HELD.

REFERENCE

ENVIRONMENT REPORTER, PUBLISHED BY THE BUREAU OF NATIONAL AFFAIRS, INC., VOL. 4, NO. 23, OCT. 5, 1973, PP 889-890

CASE	CAROLINA ACTION VS. SIMON, NO. 75-1253
LOCATION	DURHAM, NORTH CAROLINA
DATE & STATUS	REINSTATED ON 25 JUNE, 1975
PROJECT TYPE	JUDICIAL BUILDING AND CITY HALL
ISSUE(S)	NEPA APPLICABILITY, REVENUE SHARING
SCOPE	THE U.S. COURT OF APPEALS FOR THE FOURTH CIRCUIT RULED THAT A PROJECT BUILT WITH FEDERAL REVENUE SHARING FUNDS IS NOT SUBJECT TO NEPA. THE CASE CONCERNED THE CONSTRUCTION IN DURHAM OF A COUNTY JUDICIAL BUILDING AND CITY HALL. THE U.S. DISTRICT COURT FOR THE MIDDLE DISTRICT OF NORTH CAROLINA RULED THAT NEPA DOES NOT APPLY TO A PROJECT IN WHICH THE ONLY FEDERAL PARTICIPATION IS THE DISTRIBUTION OF REVENUE SHARING FUNDS TO AID A COMMUNITY IN FINANCING A LOCAL PROJECT. IN A BRIEF PER CURIAM OPINION, THE APPEALS COURT AFFIRMED THE RULING OF THE DISTRICT COURT.
REFERENCE	ENVIRONMENT REPORTER, PUBLISHED BY THE BUREAU OF NATIONAL AFFAIRS, INC., VOL. 6, NO.10, JULY 4, 1975, P 414

CASE	CEDAR-RIVERSIDE VS. HILLS, NO. 4-73-592
LOCATION	MINNEAPOLIS, MINNESOTA
DATE & STATUS	ENJOINED ON 29 MARCH, 1976
PROJECT TYPE	NEW TOWN
ISSUE(S)	IMPACT STATEMENT ADEQUACY, ECONOMIC FEASIBILITY, HIGH DENSITY DEVELOPMENT

SCOPE

THE U.S. DISTRICT COURT FOR THE DISTRICT OF MINNESOTA ENJOINED ALL FURTHER WORK ON A MINNEAPOLIS "NEW TOWN IN TOWN" PROJECT UNTIL AN ADEQUATE NEPA IMPACT STATEMENT IS PREPARED.

JUDGE MILES LORD APPROVED A SPECIAL MASTER'S REPORT CONCLUDING THAT THE IMPACT STATEMENT PREPARED ON THE CEDAR-RIVERSIDE PROJECT FAILED TO SATISFY NEPA AND THE MINNESOTA ENVIRONMENTAL POLICY ACT.

THE CEDAR-RIVERSIDE PROJECT IS PART OF THE HUD'S NEW COMMUNITIES PROGRAM. THE PROJECT'S DEVELOPERS CONTEMPLATE A 10-STAGE DEVELOPMENT PROGRAM THAT WILL CULMINATE IN 1991 WITH A PROJECTED POPULATION FOR THE NEW TOWN OF BETWEEN 25,000 AND 30,000. THE FIRST STAGE OF THE PROJECT, CONSISTING OF APPROXIMATELY 1,300 DWELLING UNITS, HAS BEEN COMPLETED.

THE IMPACT STATEMENT PREPARED BY HUD WAS CHALLENGED BY THE CEDAR-RIVERSIDE ENVIRONMENTAL DEFENSE FUND, WHICH IS COMPOSED OF RESIDENTS OF THE NEW TOWN, AND BY OTHER LOCAL CITIZENS GROUPS. JUDGE LORD APPOINTED SPECIAL MASTER EDWARD J. PARKER TO REVIEW THE ADEQUACY OF THE STATEMENT.

THE IMPACT STATEMENT IS INADEQUATE, PARKER SAID, BECAUSE IT FAILS TO PROVIDE SUFFICIENT DATA ON THE PROJECT'S ECONOMIC FEASIBILITY AND BECAUSE IT FAILS TO ASSESS THE IMPACT OF HIGH DENSITY DEVELOPMENT AS PROPOSED FOR CEDAR-RIVERSIDE.

THE IMPACT STATEMENT, PARKER SAID, FAILS TO ASSESS ADEQUATELY ALTERNATIVES TO HIGH-DENSITY DEVELOPMENT AND GIVES NO CONSIDERATION AT ALL TO THE "MOST PROBABLE" ALTERNATIVE, WHICH IS "THE INABILITY TO CARRY FORWARD THE PROJECT AS PLANNED FOR FINANCIAL REASONS." THE STATEMENT, ACCORDING TO PARKER, "DEMONSTRATES A PATTERN SHOWING BIAS ON THE PART OF HUD TOWARD JUSTIFYING A PREVIOUSLY MADE DECISION."

REFERENCE

ENVIRONMENT REPORTER, V.6, NO. 51, APR. 16, 1976, P 2109

CASE	CHELSEA NEIGHBORHOOD ASSOCIATIONS VS. U.S. POSTAL SERVICE, NO. 75-6005
LOCATION	MANHATTEN, NEW YORK
DATE & STATUS	ENJOINED ON 30 APRIL, 1975
PROJECT TYPE	VEHICLE MAINTENANCE FACILITY HOUSING PROJECT
ISSUE(S)	NEPA APPLICABILITY TO POSTAL SERVICE

SCOPE

THE U.S. POSTAL SERVICE IS NOT EXEMPTED FROM THE REQUIREMENTS OF NEPA, THE U.S. COURT OF APPEALS FOR THE SECOND CIRCUIT HELD.

THE CASE INVOLVED A CHALLENGE BY SEVERAL CITIZEN ASSOCIATIONS TO A PROPOSED VEHICLE MAINTENANCE FACILITY HOUSING PROJECT IN LOWER MANHATTEN. THE POSTAL SERVICE WAS TO CONSTRUCT THE MAINTENANCE FACILITY, WITH THE AIR SPACE GRANTED TO NEW YORK CITY FOR PUBLIC HOUSING.

THE U.S. POSTAL SERVICE PREPARED A NEPA IMPACT STATEMENT ON THE PROJECT, WHICH IT CONCEDED WOULD HAVE A MAJOR IMPACT ON THE ENVIRONMENT, BUT AT THE SAME TIME ARGUED THAT IT WAS EXEMPTED FROM NEPA REQUIREMENTS BY SECTION 410 OF THE POSTAL REORGANIZATION ACT OF 1970, 39 U.S.C. SECTION 410, WHICH PROVIDES THAT NO LAW DEALING WITH FEDERAL CONTRACTS OR PROPERTY APPLIES TO THE U.S. POSTAL SERVICE.

IN A DECISION BY JUDGE WILFRED FEINBERG, THE COURT HELD THAT THE BROAD ENVIRONMENTAL MANDATE OF NEPA WAS APPLICABLE TO THE POSTAL SERVICE. THE COURT CONTRASTED THE SPECIFIC CONCERNS OF SECTION 410 AND THE BROAD POLICY MANDATE OF NEPA: "THE POSTAL REORGANIZATION ACT IS MANAGERIAL; NEPA IS POLICY ORIENTATED. IN FACT, INSOFAR AS THE FORMER ACT WAS DESIGNED TO ENHANCE DAY-TO-DAY BUSINESS DISCRETION, NEPA IN NO WAY INTERFERES, NEPA DOES NOT REQUIRE SPECIFIC RESULTS IN PARTICULAR SITUATIONS. IT MERELY ESTABLISHES A 'REORDERING OF PRIORITIES, SO THAT ENVIRONMENTAL COSTS AND BENEFITS WILL ASSUME THEIR PROPER PLACE ALONG WITH OTHER CONSIDERATIONS.'"

ONCE DECIDING THAT NEPA APPLIED TO THE POSTAL SERVICE, THE COURT THEN FOUND THAT THE SERVICE'S IMPACT STATEMENT ON THE PROJECT FAILED TO MEET NEPA REQUIREMENTS. THE COURT SAID THAT THE POSTAL SERVICE'S STATEMENT DID NOT CONTAIN A COMPREHENSIVE ANALYSIS OF ENVIRONMENTAL AND SOCIAL IMPACTS OF THE HOUSING ASPECTS OF THE PROJECT. THE POSTAL SERVICE HAD ARGUED THAT IT SHOULD NOT BE REQUIRED TO ASSESS THE HOUSING ASPECTS, SINCE THAT PORTION OF THE PROJECT WAS THE RESPONSIBILITY OF NEW YORK CITY.

THE COURT RULED, HOWEVER, THAT IF THE POTENTIAL IMPACT OF THE HOUSING IS NOT CONSIDERED BEFORE THE VEHICLE MAINTENANCE FACILITY IS CONSTRUCTED, "IT WILL BE TOO LATE TO REASSESS THE PROJECT AS A WHOLE NO MATTER WHAT IS SHOWN BY A LATER EIS FOR THE HOUSING PREPARED BY ANOTHER AGENCY."

REFERENCE

ENVIRONMENT REPORTER, PUBLISHED BY THE BUREAU OF NATIONAL AFFAIRS, INC., VOL. 6, NO. 3, MAY 16, 1975, PP 138-139

CASE	CHICK VS. HILLS, NO. 75-1274
LOCATION	BOSTON, MASSACHUSETTS
DATE & STATUS	REINSTATED ON 22 JANUARY, 1976
PROJECT TYPE	HOUSING PROJECT
ISSUE(S)	SEPARATION OF PROJECT FROM TOTAL DEVELOPMENT PLAN, NEPA CONSIDERATIONS
SCOPE	A FEDERAL APPEALS COURT REFUSED TO OVERTURN A LOWER COURT RULING THAT HUD'S REVIEW OF A SINGLE HOUSING PROJECT DID NOT VIOLATE THE NEPA. A GROUP OF BOSTON RESIDENTS ORIGINALLY HAD CHARGED THAT HUD HAD UNLAWFULLY SEGMENTED THE PROJECT FROM THE AREA'S OVERALL DEVELOPMENT PLAN, AND THEY SAID THE SPECIAL ENVIRONMENTAL CLEARANCE PREPARED BY HUD WAS INADEQUATE. IN A PER CURIAM OPINION, THE U.S. COURT OF APPEALS FOR THE FIRST CIRCUIT SAID THE EVIDENCE SUPPORTED THE TRIAL COURT'S DETERMINATION THAT HUD HAD NOT RECOGNIZED AS A REDEVELOPMENT UNIT THE PARCEL OF LAND IN WHICH THE PROJECT WAS INCLUDED, ALSO REFUSING TO OVERTURN THE LOWER COURT'S FINDING THAT THERE WAS NO EXPECTATION OF ADDITIONAL HUD FINANCING OF OTHER CONSTRUCTION IN THE PARCEL.
REFERENCE	ENVIRONMENT REPORTER, VOL. 6, NO. 41, FEB. 6, 1976, P 1728

CASE	CITIZENS' ASSOCIATION OF GEORGETOWN VS. WASHINGTON, NO. 1944-73
LOCATION	GEORGETOWN, WASHINGTON, D.C.
DATE & STATUS	REINSTATED ON 8 AUGUST, 1974
PROJECT TYPE	BUILDING COMPLEX OF HOTEL, RESTAURANT, OFFICES AND SHOPPING CENTER
ISSUE(S)	AIR STANDARDS

SCOPE

JUDGE CHARLES R. RICHEY OF THE U.S. DISTRICT COURT FOR THE DISTRICT OF COLUMBIA DISMISSED A SUIT TO BLOCK CONSTRUCTION OF THE INITIAL PART OF A PROPOSED $80 MILLION BUILDING COMPLEX ALONG THE GEORGETOWN WATERFRONT IN WASHINGTON D.C.

THE SUIT WAS BROUGHT BY THE COMMITTEE OF 100 ON THE FEDERAL CITY AND THE CITIZENS ASSOCIATION OF GEORGETOWN WHO CHARGED THAT COMPLETION OF TWO BUILDINGS PRESENTLY UNDER CONSTRUCTION IN THE AREA WILL CAUSE A VIOLATION OF THE NATIONAL AMBIENT AIR STANDARDS IN 1977. THE BUILDINGS ARE PART OF A PROPOSED RESTAURANT, HOTEL, OFFICE, AND SHOPPING CENTER COMPLEX TO BE BUILT OVER THE NEXT FIVE YEARS BY AFFILIATES OF INLAND STEEL CORPORATION.

RICHEY RULED THAT THE DISTRICT COURT HAS JURISDICTION TO HEAR THE SUIT BY VIRTUE OF SECTION 304 OF THE CLEAN AIR ACT, AND THAT JURISDICTION ALSO LIES UNDER THE FEDERAL QUESTION STATUTE, SINCE A SUBSTANTIAL FEDERAL QUESTION EXISTS AS TO WHETHER THE ACT "IN ITS ENTIRETY" REQUIRES REVIEW OF THE DEVELOPMENT FOR POSSIBLE VIOLATION OF NATIONAL AIR STANDARDS.

AS IN HIS INITIAL OPINION DENYING PLAINTIFFS' MOTION FOR A PRELIMINARY INJUNCTION, RICHEY NOTED THAT THE DISTRICT OF COLUMBIA REGULATIONS GOVERNING INDIRECT SOURCES SUCH AS THE BUILDINGS UNDER CONSTRUCTION APPLY ONLY TO FACILITIES WHOSE CONSTRUCTION COMMENCED AFTER NOVEMBER 12, 1973. THE BUILDINGS IN QUESTION WERE BEGUN BEFORE THAT DATE, AND HENCE ARE NOT GOVERNED BY THE INDIRECT SOURCE REGULATIONS.

RICHEY RULED THAT THE PLAINTIFFS HAD FAILED TO DEMONSTRATE THAT CONSTRUCTION OF THE BUILDINGS WOULD LEAD TO A VIOLATION OF NATIONAL AIR STANDARDS. THE EVIDENCE ABOUT THE AIR QUALITY IN THE GEORGETOWN AREA IN 1977, HE SAID, IS "SPECULATIVE, CONFLICTING, AND INADEQUATE TO SUPPORT A JUDGEMENT." HE ALSO NOTED THAT GASOLINE SHORTAGES, FEDERAL AUTOMOBILE EMISSION STANDARDS, FEDERAL AND LOCAL LIMITATIONS ON PARKING, AND OTHER FACTORS COULD LEAD TO A SUBSTANTIAL REDUCTION OF POLLUTION IN THE GEORGETOWN AREA.

REFERENCE

ENVIRONMENT REPORTER, PUBLISHED BY THE BUREAU OF NATIONAL AFFAIRS, INC., VOL. 5, NO. 16, AUGUST, 1974, P 480

FOR THE FULL TEXT OF THE JUDGES OPINION, SEE:
ENVIRONMENT REPORTER CASES, VOL. 6, P 1166.

CASE	CITY OF BOSTON VS. MASS. PORT OF AUTHORITY, NO. 7844
LOCATION	BOSTON, MASSACHUSETTS
DATE & STATUS	REMANDED FEBRUARY 27, 1974
PROJECT TYPE	NEW TERMINAL AND PARKING GARAGE
ISSUE(S)	AIR POLLUTION REGULATIONS, EIS ADEQUACY
SCOPE	PLAINTIFFS SOUGHT REVERSAL OF A LOWER COURT DECISION RULING THAT DEFENDANTS ARE EXEMPT FROM STATE AIR POLLUTION REGULATIONS AND HAVE ADEQUATELY REVIEWED THE ENVIRONMENTAL IMPACT OF CONSTRUCTION OF A PROPOSED NEW TERMINAL AND 5-STORY PARKING GARAGE WITH 2,700 PARKING SPACES AT LOGAN AIRPORT. THE PLAINTIFF REQUESTS INJUNCTION RELIEF. THE APPEALS COURT AFFIRMED THE LOWER COURT RULING THAT DEFENDANTS HAVE COMPLIED WITH THE EFFECTIVE PROVISIONS FOR AN IMPACT STATEMENT WITH RESPECT TO THE PLANNED CONSTRUCTION. HOWEVER, THE COURT REVERSED THE LOWER COURT'S HOLDING THAT THE CITY HEALTH DEPARTMENT'S AIR POLLUTION CONTROL REGULATIONS, WHICH ARE DESIGNED TO ALLEVIATE A PERVASIVE PUBLIC HEALTH PROBLEM, AND WHICH, BY VIRTUE OF THE STATUTES AUTHORIZING IT, ARE UNIVERSALLY APPLICABLE TO ALL PRIVATE AND PUBLIC ENTITIES. ACCORDING TO THE STATE PUBLIC HEALTH DEPARTMENT, ITS REGULATIONS PROHIBIT THE CONSTRUCTION OF ANY FACILITY WHICH MAY CONTRIBUTE TO A CONDITION OF AIR POLLUTION UNTIL THE PLANS ARE REVIEWED AND APPROVED BY THE DEPARTMENT. THE DEFENDANTS ARGUED THAT THE CONSTRUCTION DOES NOT CONSTITUTE A VIOLATION OF A STATUTE BECAUSE THERE IS NO ALLEGATION THAT THERE IS ENVIRONMENTAL DAMAGE. THE COURT RULED THAT THE CONTROVERSY DOES NOT WARRANT A DISMISSAL, AND THE CASE IS TO BE REMANDED TO THE LOWER COURT FOR FURTHER PROCEEDINGS.
REFERENCE	ENVIRONMENTAL LAW REPORTER, VOL. 4, P 20314

CASE	CITY OF DALLAS VS. CROWNRICH, NO. 506 S.W. 2D 654
LOCATION	TEXAS
DATE & STATUS	ENJOINED 1974
PROJECT TYPE	HIGH-RISE APARTMENT BUILDING
ISSUE(S)	BUILDING PERMIT ISSUANCE, ZONING CODES
SCOPE	THE COURT HELD THAT WITHHOLDING THE PERMIT PURSUANT TO A CITY COUNCIL RESOLUTION WAS A VALID EXERCISE OF POLICE POWER. THE COURT ASSERTED THAT IN THE INTERIM BETWEEN THE PROPOSING AND ADOPTING OF A ZONING ORDINANCE, THE CITY COULD TAKE REASONABLE MEASURES TEMPORARILY TO PROTECT THE PUBLIC INTEREST AND WELFARE. POLICE POWER HAD BEEN USED TO PROHIBIT THE CONSTRUCTION OF A HIGH-RISE APARTMENT BUILDING THAT WOULD HAVE HAD AN INCONGRUOUS EFFECT ON AN AREA WHICH WAS THE SUBJECT OF A PROPOSED HISTORIC DISTRICT. A BUILDING PERMIT TO CONSTRUCT AN APARTMENT BUILDING WAS WITHHELD BY THE CITY UNTIL THE MATTER OF REZONING THE AREA, IN WHICH THE APARTMENT BUILDING WAS TO BE BUILT, HAD BEEN RESOLVED. AT THE TIME OF THE PERMIT REQUEST, THE AREA WAS ZONED TO ALLOW SUCH CONSTRUCTION. HOWEVER, A MONTH BEFORE THE REQUEST, ON RECOMMENDATION BY THE CITY PLANNING COMMISSION, THE CITY COUNCIL HAD BEGUN TO CONSIDER SERIOUSLY AN HISTORIC PRESERVATION ORDINANCE WHICH WOULD REZONE THAT AREA INTO AN HISTORIC DISTRICT AND WOULD PROHIBIT THE PROPOSED CONSTRUCTION.
REFERENCE	WAKE FOREST LAW REVIEW, VOL. 12, 1976, P 256

CASE	CITY OF ROCHESTER VS. U.S. POSTAL SERVICE, NO. 76-6065
LOCATION	ROCHESTER, NEW YORK
DATE & STATUS	REINSTATED SEPTEMBER 3, 1976
PROJECT TYPE	NEW POSTAL FACILITY
ISSUE(S)	NEPA APPLICABILITY, TRAFFIC, URBAN DECAY
SCOPE	THE COURT HELD THAT THE POSTAL SERVICE ERRED IN CONCLUDING THAT CONSTRUCTION OF A NEW POSTAL FACILITY IN A SUBURB OF ROCHESTER, NEW YORK, DID NOT REQUIRE PREPARATION OF A FULL-SCALE NEPA IMPACT STATEMENT. THE SERVICE'S ENVIRONMENTAL ASSESSMENT FOR THE PROJECT NEGLECTED TO CONSIDER A NUMBER OF POSSIBLE MAJOR ENVIRONMENTAL EFFECTS, INCLUDING INCREASED COMMUTER TRAFFIC AND URBAN DECAY CAUSED BY THE TRANSFER OF 1,400 EMPLOYEES FROM THE EXISTING DOWNTOWN POST OFFICE TO THE NEW FACILITY. THUS, THE COURT GRANTED THE REGIONAL PLANNING BOARD OF THE CITY OF ROCHESTER UNDER THE INTERGOVERNMENTAL COOPERATIVE ACT STANDING TO SEEK REVIEW OF THE FEDERAL PROJECT WHICH DOES NOT COMPORT WITH THE REGIONAL PLAN. THE COURT RULED THAT THE POSTAL SERVICE IS NOT IMMUNE FROM THE STATUTE. HOWEVER, THE COURT DECLINED TO ENJOIN CONSTRUCTION OF THE FACILITY WHICH IS NOW 35% COMPLETE BECAUSE OF THE ECONOMIC WASTE THAT WOULD ENTAIL. THE COURT DID ENJOIN THE TRANSFER OF ACTIVITIES OR EMPLOYEES FROM THE OLD POST OFFICE TO THE NEW FACILITY PENDING PREPARATION OF AN ADEQUATE NEPA IMPACT STATEMENT.
REFERENCE	ENVIRONMENTAL LAW REPORTER, VOL. 6, P 20723

CASE	CLINTON COMMUNITY HOSPITAL CORPORATION VS. SOUTHERN MARYLAND MEDICAL CENTER, NO. 73-981-HM
LOCATION	MARYLAND
DATE & STATUS	REINSTATED APRIL 16, 1974
PROJECT TYPE	HOSPITAL CONSTRUCTION
ISSUE(S)	NEPA APPLICABILITY, AESTHETICS, POSSIBILITY OF PHYSICAL INJURY
SCOPE	THE COURT RULED THAT THE PLAINTIFF'S SUIT FOR AN INJUNCTION AGAINST CONSTRUCTION OF DEFENDANT'S COMPETING HOSPITAL FAILS TO ASSERT AN INJURY IN FACT WITHIN THE ZONE OF INTERESTS OF THE NEPA. THUS, PLAINTIFF LACKED STANDING TO SEEK RELIEF. THE COURT HELD THAT AS A CORPORATION, PLAINTIFF CANNOT ASSERT AN INJURY TO ITS AESTHETIC ENJOYMENT OF THE ENVIRONMENT, AND THEREFORE CANNOT DEMONSTRATE THE REQUISITE INJURY IN FACT TO GAIN STANDING. JUDGE MURRAY, WRITING FOR THE COURT, DECLARED THAT PLAINTIFF'S CLAIM ON BEHALF OF THE EMPLOYEES AND CONSUMERS OF DEFENDANT HOSPITAL SERVICES DID NOT CONFER STANDING, FOR TO EXTEND THE REQUIREMENTS OF NEPA TO EVERY ACTION AFFECTING PEOPLE WOULD BE AN EGREGIOUS AND UNWARRANTED EXTENSION OF THE SCOPE OF THE STATUTE. IN OTHER WORDS, AGENCIES MUST EXAMINE THE IMPACT OF THE ACTION ON THE GENERAL ENVIRONMENT, NOT THE IMPACT OF THE ENVIRONMENT UPON THE ACTION. PLAINTIFF'S ASSERTION THAT POTENTIAL DAMAGE TO THE DEFENDANT HOSPITAL UNDER AN AIRCRAFT FLIGHT PATH RESULTING FROM ACCIDENTS INVOLVING AIRCRAFT WAS INAPPROPRIATE, AS FOUND BY THE COURT. ALSO, THE COURT FOUND THEIR CLAIMS UNDER THE COMPREHENSIVE HEALTH PLANNING ACT, THE NATIONAL HOUSING ACT, THE PUBLIC HEALTH SERVICE ACT, AND THE NOISE CONTROL ACT TO BE WITHOUT MERIT.
REFERENCE	ENVIRONMENTAL LAW REPORTER, VOL. 4, PP 20670-20674

CASE	COALITION FOR THE ENVIRONMENT VS. VOLPE, NOS. 72-1769, 72-1770
LOCATION	ST. LOUIS COUNTY, MISSOURI
DATE & STATUS	ENJOINED ON 31 JULY, 1974
PROJECT TYPE	DEVELOPMENT OF RESIDENTIAL, COMMERCIAL AND INDUSTRIAL COMMUNITY
ISSUE(S)	OPEN SPACE, NOISE AND AIR POLLUTION, POPULATION DENSITY, LAND USE PLANNING

SCOPE

A FEDERAL DISTRICT COURT DECISION THAT DISMISSED, DUE TO LACK OF STANDING, A SUIT CHALLENGING THE CONSTRUCTION OF THE PROPOSED "EARTH CITY" DEVELOPMENT PROJECT IN MISSOURT WAS REVERSED BY THE U.S. COURT OF APPEALS FOR THE EIGHTH CIRCUIT.

CONSTRUCTION OF EARTH CITY, A PLANNED RESIDENTIAL, COMMERCIAL, AND INDUSTRIAL COMMUNITY , WAS CHALLENGED BY SEVERAL ENVIRONMENTAL GROUPS AND INDIVIDUALS. THEY ALLEGED HARM CAUSED BY THE LOSS OF OPEN SPACE, BY THE INCREASED NOISE AND AIR POLLUTION FROM THE DEVELOPMENT, AND BY INCREASED POPULATION DENSITY IN THE COUNTY.

THE U.S. DISTRICT COURT FOR THE EASTERN DISTRICT OF MISSOURI RULED THAT THE PLAINTIFFS HAD FAILED TO SATISFY THE "INJURY IN FACT" PORTION OF TEE STANDING TEST ENUNCIATED IN U.S. VS. SCRAP. THE ONLY INJURY ALLEGED BY THE PLAINTIFFS, THE DISTRICT COURT SAID, "IS THE APPARENT FACT THAT THE ACTIONS OF THE DEFENDENTS ARE PERSONALLY DISPLEASING" DUE TO THE PARTIES' DIFFERING PHILOSOPHIES OF LAND USE PLANNING.

THE APPEALS COURT DISAGREED. THE PLAINTIFFS' ALLEGATIONS, THE COURT SAID, "CONSTITUTE MORE THAN SIMPLE ASSERTIONS OF DISTASTE OR DISPLEASURE; THEY ARE STATEMENTS OF SPECIFIC INJURY EXPERIENCED BY ASCERTAINABLE INDIVIDUALS WHO RESIDE NEAR OR PASS THROUGH THE AFFECTED AREA."

THE APPELLATE COURT RULING INVOLVED ONLY WHETHER PLAINTIFFS MET THE "INJURY IN FACT" TEST, AND DID NOT ADDRESS THE QUESTION OF WHETHER PLAINTIFFS SATISFIED THE SECOND PORTION OF THE STANDING TEST, THE NEED TO SHOW THAT THE INJURY WAS ARGUABLY WITHIN THE ZONE OF INTEREST TO BE PROTECTED.

THE APPELLATE COURT ALSO DID NOT REACH THE MERITS OF THE CASE, WHICH INVOLVES PLAINTIFFS' ASSERTIONS THAT APPROVAL OF A PROPOSED LEVEE BY THE CORPS OF ENGINEERS AND THE APPROVAL OF A PROPOSED INTERSTATE HIGHWAY INTERCHANGE BY THE DEPARTMENT OF TRANSPORTATION WERE CRUCIAL TO THE DECISION TO PROCEED WITH THE EARTH CITY PROJECT AND SHOULD HAVE BEEN ACCOMPANIED BY NEPA IMPACT STATEMENTS.

REFERENCE

ENVIRONMENT REPORTER, PUBLISHED BY THE BUREAU OF NATIONAL AFFAIRS, INC., VOL.5, NO. 15, AUG. 9, 1974, PP 445-446

FOR THE FULL TEXT OF THE JUDGE'S OPINION SEE:
ENVIRONMENT REPORTER CASES, VOL. 4, P 1509.

FOR THE FULL TEXT OF THE PRECIDENT, SEE:
ENVIRONMENT REPORTER CASES, VOL. 5, P 1449

CASE	COMMITTEE TO SAVE THE SOUTH GREEN VS. HILLS, NO. H-76-328
LOCATION	CONNECTICUT
DATE & STATUS	ENJOINED NOVEMBER 5, 1976
PROJECT TYPE	URBAN RENEWAL CONSTRUCTION PROJECT
ISSUE(S)	APPLICABILITY OF NEPA AND NHPA
SCOPE	THE COURT ENJOINED DEMOLITION OF AN HISTORICAL HOUSE AND FURTHER CONSTRUCTION ON AN URBAN RENEWAL PROJECT PENDING TRIAL ON THE MERITS OF PLAINTIFF'S CLAIM THAT THE PROCEDURAL REQUIREMENTS OF NEPA AND NHPA ARE APPLICABLE TO THE PROJECT. THE SOUTH GREEN DISTRICT WAS PLACED ON THE NATIONAL REGISTER OF HISTORIC PLACES IN 1975, THE CONTRACT WAS EXECUTED IN 1970. NO EIS OF THE HIGHWAY PROJECT WAS MADE, AND THE LOCAL REDEVELOPMENT AGENCY FOUND THE REHABILITATION COSTS OF THE MATHER HOUSE PROHIBITIVE. IN GRANTING PLAINTIFF'S MOTION FOR AN INJUNCTION, THE COURT REJECTED EVERY DEFENSE RAISED BY FEDERAL AND LOCAL URBAN RENEWAL OFFICIALS. THE COURT FOUND THAT PLAINTIFFS HAD PROVEN PROBABLE SUCCESS ON THE MERITS SUFFICIENT TO JUSTIFY THE PRELIMINARY INJUNCTION. FIRST, THE DEPARTMENT OF HUD SHOULD HAVE FOLLOWED THE CONSULTATION PROCEDURES OF NHPA. AND SECOND, THE PLAINTIFFS AND PUBLIC INTEREST IN PRESERVING THE ENVIRONMENT WOULD BE IRREPARABLY INJURED IF THE PROJECT WAS PERMITTED TO PROCEED, WHILE THE HARMS THREATENED TO DEFENDANTS DID NOT OUTWEIGH THE POSSIBLE BENEFITS OF A DELAY PENDING FULL TRIAL ON THE MERITS.
REFERENCE	ENVIRONMENTAL LAW REPORTER, VOL. 7, P 20061

CASE	CONCERNED ABOUT TRIDENT VS. RUMSFELD, NOS. 75-1515 AND 75-2053
LOCATION	BANGOR, WASHINGTON
DATE & STATUS	DISMISSED OCTOBER 13, 1976
PROJECT TYPE	NAVY ATOMIC SUBMARINE FACILITY CONSTRUCTION
ISSUE(S)	NEPA COMPLIANCE, EIS ADEQUACY

SCOPE

THE U.S. COURT OF APPEALS FOR THE DISTRICT OF COLUMBIA CIRCUIT RULED THAT THE U.S. NAVY MUST REVISE IN PART ITS IMPACT STATEMENT ON ITS TRIDENT ATOMIC MISSILE SUBMARINE SYSTEM. THE COURT, HOWEVER, DID NOT ENJOIN WORK ON THE TRIDENT SITE AT BANGOR, WASHINGTON, PENDING NEPA COMPLIANCE.

THE APPEALS COURT REVERSED, IN PART, A LOWER COURT RULING, BY STATING THAT NEPA REQUIRES A MORE EXTENSIVE DISCUSSION OF ALTERNATIVES TO THE BANGOR SITE.

THE COURT IN A PER CURIAM OPINION SAID ALSO THAT THE STATEMENT SHOULD HAVE DISCUSSED POSSIBLE IMPACTS ON THE CHOSEN SITE PAST THE PROJECT'S PROJECTED 1981 COMPLETION DATE. WHILE THE NAVY "NEED NOT, AND INDEED, MAY NOT BE ABLE TO FORECAST THE EFFECTS OF TRIDENT AFTER 1981 IN THE SAME DETAIL OR WITH THE SAME DEGREE OF ACCURACY AS IT HAS DONE FOR THE PERIOD PRIOR TO 1981", THE COURT SAID, "IT IS IMPERATIVE THAT IT MAKE A REASONABLE EFFORT TO DISCERN WHAT THE EFFECTS OF TRIDENT'S FUTURE OPERATION WILL BE".

THE COURT REJECTED THE NAVY'S CONTENTION THAT ITS ACTIONS ARE NOT GOVERNED BY NEPA, SAYING THAT ARGUMENT REPRESENTED "A FLAGRANT ATTEMPT TO EXEMPT FROM THE MANDATES OF NEPA ALL SUCH MILITARY ACTIONS UNDER THE OVERUSED RUBRIC OF 'NATIONAL DEFENSE'".

THE COURT ALSO REJECTED THE ARGUMENTS THAT A SEPARATE "PROGRAM" IS WAS REQUIRED AND THAT THE STATEMENT FAILED TO ADEQUATELY ASSESS ENVIRONMENTAL IMPACTS OUTSIDE OF KITSAP COUNTY, WASHINGTON, THE IMPACTS OF POSSIBLE DEPLOYMENT OF MORE THAN TEN SUBMARINES AND THE EARLY TERMINATION OF THE TRIDENT PROGRAM, AND THE PROGRAM'S FISCAL IMPACTS.

REFERENCE

ENVIRONMENT REPORTER, PUBLISHED BY THE BUREAU OF NATIONAL AFFAIRS, INC., VOL. 7, NO. 25, OCTOBER 22, 1976, P. 924

CASE	CROSLEY BUILDING CORPORATION VS. SAMPSON, NO. 75-1043
LOCATION	JACKSON, MISSISSIPPI
DATE & STATUS	REINSTATED ON 29 SEPT., 1975
PROJECT TYPE	FEDERAL OFFICE BUILDING
ISSUE(S)	NEPA REQUIREMENTS, LAND ACQUISITION AND LEASING PROCEDURE
SCOPE	A FEDERAL DISTRICT COURT DENIED A MOTION FOR A PRELIMINARY INJUNCTION TO PREVENT THE GENERAL SERVICES ADMINISTRATION FROM ENTERING INTO A CONTRACT FOR LAND ACQUISITION AND CONSTRUCTION OF A PROPOSED FEDERAL OFFICE BUILDING. THE SUIT WAS BROUGHT BY TWO CORPORATIONS WHICH ALLEGED GSA VIOLATED NEPA BY NOT PRESENTING AN ENVIRONMENTAL IMPACT STATEMENT TO THE HOUSE AND SENATE PUBLIC WORKS COMMITTEES WHEN IT PRESENTED A LEASE PROSPECTUS TO THE COMMITTEES FOR THEIR APPROVAL PURSUANT TO THE PUBLIC BUILDINGS ACT. JUDGE JOHN H. PRATT RULED THAT NEPA DOES NOT REQUIRE GSA TO FILE AN IMPACT STATEMENT WITH ITS LEASE PROSPECTUS TO THE COMMITTEES. THE SUBMISSION OF THE PROSPECTUS BY GSA WAS NOT A PROPOSAL FOR LEGISLATION WITHIN THE INTENDMENT OF NEPA, ALSO NOTING THAT WHEN GSA SUBMITTED THE LEASE PROSPECTUS TO THE COMMITTEES, THE PROPOSAL WAS NOT SUFFICIENTLY DEFINITE SO AS TO REQUIRE AN IMPACT STATEMENT. "GSA HAD MADE NO IRREVOCABLE COMMITMENTS WITH REGARD TO THE PROPOSAL FOR A FEDERAL OFFICE BUILDING, AND UNDER LEASE PROCEDURES, NO COMMITMENT TO LEASE COULD BY MADE UNTIL, IN ACCORDANCE WITH GSA PROCEDURE, GSA RECEIVED PROPOSALS TO LEASE FROM OFFERERS PURSUANT TO FORMAL SOLICITATION OF SUCH PROPOSALS," THE COURT SAID.
REFERENCE	ENVIRONMENT REPORTER, VOL. 6, NO. 25, OCT. 17, 1975, P 1013

CASE	CUMMINGTON PRESERVATION COMMITTEE VS. FAA, NO. 75-1198
LOCATION	CUMMINGTON, MASSACHUSETTS
DATE & STATUS	REINSTATED ON 23 OCTOBER, 1975
PROJECT TYPE	RADAR FACILITY
ISSUE(S)	IMPACT STATEMENT ADEQUACY, ALTERNATIVE SITES

SCOPE

THE U.S. COURT OF APPEALS FOR THE FIRST CIRCUIT UPHELD A FEDERAL AVIATION ADMINISTRATION ENVIRONMENTAL IMPACT STATEMENT ON A PROPOSED RADAR FACILITY.

THE CITIZENS GROUP SUED TO ENJOIN FAA FROM CONSTRUCTING THE FACILITY ON GROUNDS THAT THE IMPACT STATEMENT VIOLATED NEPA, DISTRICT JUDGE FRANK H. FREEDMAN RULING IN FAVOR OF FAA.

THE APPELLATE COURT, IN AN OPINION BY JUDGE EDWARD M. MCENTEE, SAID THAT ALTHOUGH ALTERNATIVE LOCATIONS WITHIN THE TOWNS OF CUMMINGTON AND WORTHINGTON DISCUSSED IN THE IMPACT STATEMENT MIGHT HAVE BEEN MORE DETAILED, "THE ADVANTAGES AND DISAGVANTAGES WERE DISCUSSED WITH SUFFICIENT INFORMATION PRESENTED TO PERMIT THE DECISIONMAKERS INTELLIGENTLY TO CHOOSE FROM AMONG THE SITES."

ALTHOUGH THE IMPACT STATEMENT FAILED TO DISCUSS ANY ALTERNATIVE LOCATIONS OUTSIDE THOSE TOWNS, THE COURT SAID A STATEMENT "NEED NOT DISCUSS EVERY CONCEIVABLE ALTERNATIVE PLAN OF ACTION, BUT ONLY THOSE WHICH ARE REASONABLE."

THE IMPACT STATEMENT'S DISCUSSION OF WHETHER THE PROJECT WOULD INFLUENCE FUTURE DEVELOPMENT WAS "MINIMALLY ADEQUATE," THE COURT SAID. EVEN THOUGH IT DID NOT DIRECTLY EXPLAIN HOW FUTURE DEVELOPMENT WOULD BE INFLUENCED, THE STATEMENT "ENABLED THOSE WHO DID NOT HAVE A PART IN ITS COMPILATION TO UNDERSTAND AND CONSIDER MEANINGFULLY THE FACTORS INVOLVED."

REFERENCE

ENVIRONMENT REPORTER, PUBLISHED BY THE BUREAU OF NATIONAL AFFAIRS, INC., VOL.6, NO. 28, NOV. 7, 1975, P. 1229

FOR THE FULL TEXT OF THE DISTRICT COURT'S OPINION, SEE: ENVIRONMENT REPORTER CASES, VOL. 8, P 1121

CASE	DI VOSTA RENTALS, INC. VS. LEE, NO. 73-1893
LOCATION	PALM BEACH, FLORIDA
DATE & STATUS	ENJOINED NOVEMBER 14, 1973
PROJECT TYPE	APARTMENT BUILDING CONSTRUCTION
ISSUE(S)	PERMITS FOR LAND FILL
SCOPE	THE COURT RULED THAT THE CORPS OF ENGINEERS DID NOT ABUSE ITS DISCRETION IN DENYING A DEVELOPER'S APPLICATION FOR A PERMIT TO FILL SUBMERGED LAND FOR APARTMENT BUILDING CONSTRUCTION. THIS RULING REVERSED A LOWER COURT DECISION REQUIRING THE CORPS TO ISSUE THE PERMIT. THE APPEALS COURT FOUND THAT THE LOWER COURT HAD ERRED IN RULING THAT THE PROPER STANDARD FOR REVIEWING SUCH ADMINISTRATIVE DECISIONS IS THE SUBSTANTIAL EVIDENCE RULE IN THE APA RATHER THAN THE ABUSE OF DISCRETION PROVISION IN THE SAME DOCUMENT. THE COURT HEREBY ORDERED THAT THE COMPLAINT BE DISMISSED.
REFERENCE	ENVIRONMENTAL LAW REPORTER, VOL. 4, P 20005

CASE	DON'T TEAR IT DOWN, INC. VS. GSA, CIV. A, NO. 74-381
LOCATION	WASHINGTON, D.C.
DATE & STATUS	REINSTATED APRIL 23, 1975
PROJECT TYPE	NEW CONSTRUCTION
ISSUE(S)	DESTRUCTION OF BUILDINGS WITH HISTORICAL SIGNIFICANCE
SCOPE	THE COURT RULED TO DISMISS AN INJUNCTION ACTION BROUGHT TO HALT THE GENERAL SERVICES ADMINISTRATION FROM COMPLETING DESTRUCTION OF CERTAIN BUILDINGS IN WASHINGTON, D.C. THE PLAINTIFFS HAD ARGUED THAT BECAUSE THE BUILDINGS ARE HISTORICALLY SIGNIFICANT, THE ADVISORY COUNCIL ON HISTORIC PRESERVATION SHOULD FIRST HAVE BEEN CONSULTED, AS REQUIRED BY THE NATIONAL HISTORIC PRESERVATION ACT OF 1966. THE PLAINTIFFS SECURED A TEMPORARY RESTRAINING ORDER BANNING FURTHER DEMOLITION ONLY AFTER SUBSTANTIAL DAMAGE TO THE BUILDING HAD BEEN CAUSED BY THE GSA. THE DEFENDANTS SUBSEQUENTLY COMPLIED WITH THE CONSULTATION REQUIREMENTS OF THE NATIONAL HISTORIC PRESERVATION ACT OF 1966. THE DEFENDANTS THEN MOVED TO DISMISS THE CASE ON THE GROUNDS THAT THE ACTION IS NOW MOOT, SINCE REQUISITE CONSULTATION HAS BEEN COMPLIED WITH. THE COURT RULED IN FAVOR OF THE DEFENDANTS, THUS DISMISSING THE ACTION.
REFERENCE	ENVIRONMENTAL LAW REPORTER, VOL. 6, PP 20091-20093

CASE	EAST 63RD STREET ASSOCIATION VS. COLEMAN, NO. 76, CIV. 2071
LOCATION	MANHATTAN
DATE & STATUS	REINSTATED MAY 17, 1976
PROJECT TYPE	SUBWAY STATION
ISSUE(S)	EIS ADEQUACY, ECONOMIC FEASIBILITY
SCOPE	THE U.S. DISTRICT COURT FOR THE SOUTHERN DISTRICT OF NEW YORK REFUSED TO HALT CONSTRUCTION OF A MANHATTAN SUBWAY STATION FOR ALLEGED VIOLATIONS OF NEPA. THE ASSOCIATION CLAIMED THE NEPA EIS ON THE ENTIRE SUBWAY PROJECT FILED BY THE DEPARTMENT OF TRANSPORTATION, WHICH PROVIDED FINANCING, WAS DEFICIENT WITH RESPECT TO THE EFFECTS OF THE EAST 63RD STREET STATION. WRITING FOR THE COURT, THE JUDGE REJECTED THE ASSOCIATION'S ARGUMENTS THAT THE IMPACT STATEMENT FAILED TO DESCRIBE THE EXTENT OF EXCAVATION REQUIRED FOR THE STATION. THE DEPARTMENT GAVE "EXPLICIT NOTICE" OF THE EXTENT OF THE CUT AND COVER WORK AT A PUBLIC HEARING FIVE MONTHS BEFORE THE STATEMENT WAS PROMULGATED. THE COURT ALSO FOUND ADEQUATE THE STATEMENT'S REFERENCES TO "TEMPORARY DISRUPTIONS" AND OTHER CONSTRUCTION IMPACTS. ALTHOUGH A DISCUSSION OF "URBAN EFFECTS" IS REQUIRED TO SOME DEGREE, THE COURT CALLED THE ASSOCIATION'S ARGUMENTS CONCERNING CRIME AND ASSOCIATED IMPACTS "ESSENTIALLY ONLY SPECULATION". THE COURT SAID THE ASSOCIATION'S UNEXCUSED, THREE YEAR DELAY IN BRINGING SUIT, DURING WHICH TIME CONSTRUCTION HAD PROCEDED SO THAT A DISRUPTION WOULD BE ECONOMICALLY IMPRACTICAL, WEIGHED HEAVILY AGAINST GRANTING AN INJUNCTION.
REFERENCE	ENVIRONMENT REPORTER, PUBLISHED BY THE BUREAU OF NATIONAL AFFAIRS, INC., VOL. 7, NO. 4, MAY 28, 1976, P 125

CASE	EDWARDS VS. FIRST BANK OF DUNDEE, NO. 75-1655
LOCATION	
DATE & STATUS	REINSTATED ON 21 APRIL, 1976
PROJECT TYPE	NEW BANK
ISSUE(S)	DESTRUCTION OF HOUSE ON SITE
SCOPE	A FEDERAL DISTRICT COURT LACKED JURISDICTION UNDER NEPA TO ENJOIN A BANK'S DESTRUCTION OF AN OLD HOUSE, SINCE THERE WAS A TOTAL ABSENCE OF FEDERAL FUNDING OR ASSISTANCE, THE SEVENTH CIRCUIT HELD. THE HOUSE STOOD ON LAND OWNED BY THE BANK AND PROPOSED FOR USE AS A NEW BANK SITE. THE GROUP BRINGING THE SUIT CLAIMED THAT THE FEDERAL DEPOSIT INSURANCE CORPORATION ACT'S REQUIREMENT THAT THE BANK OBTAIN ITS PERMISSION BEFORE MOVING ITS OFFICES PROVIDED THE FEDERAL ACTION FOR A NEPA CLAIM. IN AN OPINION BY JUDGE WILLIAM B. EAST, THE COURT SAID THE BANK'S UNDERTAKING TO DEMOLISH THE STRUCTURE IN PREPARATION FOR PUTTING THE PROPERTY TO ANOTHER USE WAS NOT FEDERALLY FUNDED OR ASSISTED IN ANY WAY. IT NOTED ALSO THAT THE GROUP DID NOT JOIN ANY FEDERAL OFFICER AS A PARTY DEFENDANT. "IT IS MANIFEST THAT THE BANK'S PROJECT," THE COURT SAID, "IS A PRIVATE ACT OVER WHICH THE COMPTROLLER HAS NO AUTHORITY, EITHER EXPRESSED OR IMPLIED UNDER FDICA, AND THE REGULATIONS THEREUNDER, EITHER TO GRANT OR WITHHOLD THE GRANTING OF A FEDERAL LICENSE TO DEMOLISH A PRIVATELY OWNED BUILDING."
REFERENCE	ENVIRONMENT REPORTER, VOL. 6, NO. 53, APR. 30, 1976, P 1991

CASE	EDWARDS VS. FIRST BANK OF DUNDEE, NO. 393 F. SUPP. 680
LOCATION	ILLINOIS
DATE & STATUS	ENJOINED 1975
PROJECT TYPE	RECONSTRUCTION OF BANK OFFICES
ISSUE(S)	NATIONAL HISTORICAL PRESERVATION ACT APPLICABILITY, COMMUNITY NEEDS
SCOPE	THE COURT FOUND THAT THE NHPA APPLIED TO THE PARTICULAR FACTS OF THE CASE. THE PLAINTIFFS SOUGHT TO ENJOIN THE BANK FROM DEMOLISHING A HOUSE LOCATED IN AN AREA LISTED IN THE NATIONAL REGISTER OF HISTORIC PLACES, OR TO ENJOIN THE VILLAGE OF WEST DUNDEE FROM ISSUING A DEMOLITION PERMIT TO THE BANK. THE DEFENDANTS ARGUED THAT THE REQUISITE FEDERAL INVOLVEMENT WAS MISSING AND THAT THE NHPA WAS INAPPLICABLE. BECAUSE THE BANK WISHED TO RELOCATE ITS OFFICES ON THE SITE OCCUPIED BY THE HISTORIC HOUSE, DEMOLITION OF THE HOUSE WAS NECESSARY. THE COURT FOUND THAT THE BANK WAS REQUIRED UNDER THE FEDERAL DEPOSIT INSURANCE CORPORATION ACT AND REGULATIONS TO OBTAIN A PERMIT FROM THE FDIC TO ESTABLISH OR MOVE ITS MAIN BRANCH OR OFFICE. SINCE THE FDIC HAD TO CONSIDER THE COMMUNITY NEEDS SERVED BY THE RELOCATED BANK, THE COURT RULED THAT THE ACTIONS OF THE FEDERAL AGENCY, THE FDIC, CAME WITHIN THE NHPA. IN ADDITION, THE COURT STATED THAT IT HAD THE POWER TO ENJOIN NON-FEDERAL ENTITIES FROM PERFORMING ACTIVITIES IN CONTRAVENTION OF THE FEDERAL AGENCY'S RESPONSIBILITIES TO PREVENT THOSE ACTIVITIES.
REFERENCE	WAKE FOREST LAW REVIEW, VOL. 12, 1976, P 261

CASE	ELY VS. VELDE, NO. 73-1889
LOCATION	LOUISA COUNTY, VIRGINIA
DATE & STATUS	REINSTATED MAY 8, 1974
PROJECT TYPE	MEDICAL CENTER FOR PRISONERS
ISSUE(S)	COMPLIANCE WITH NEPA AND NHPA
SCOPE	THE COURT RULED THAT THE STATE OF VIRGINIA MAY NOT BUILD A MEDICAL CENTER FOR PRISONERS WITHOUT EITHER COMPLIANCE WITH THE DICTATES OF NEPA AND NATIONAL HISTORIC PRESERVATION ACT OR REIMBURSEMENT OF FEDERAL FUNDS, ORIGINALLY ALLOCATED UNDER LEAA'S BLOCK GRANT FUNDING FOR CONSTRUCTION OF THE CENTER, WHICH WERE TRANSFERRED BY THE STATE TO OTHER PROJECTS. A LOWER COURT HAD RULED THAT NEPA'S REQUIREMENTS DID NOT APPLY TO A STATE MEDICAL CENTER FOR PRISONERS ONCE THE STATE HAD INDICATED THAT IT WOULD NOT USE ANY FEDERAL FUNDS FOR CONSTRUCTION OF THE CENTER. THE APPEALS COURT, HOWEVER, FOUND THAT THE RELOCATION OF FEDERAL FUNDS IS A SUBVERSION AND CIRCUMVENTION OF ACTS OF CONGRESS. THE BLOCK GRANT FUNDING IS CONDITIONED UPON COMPLIANCE WITH FEDERAL LAW. HOWEVER, THE COURT DENIED PLAINTIFF'S REQUEST FOR AN INJUNCTION, STATING THAT AS OF YET THE MEDICAL CENTER HAD NOT BECOME AN IRREVOCABLY FEDERAL PROJECT. IN AN OPINION BY JUDGE BUTZNER, HE STATED THAT CONSTRUCTION MAY PROCEED, BUT "IF IT IS TO BE CONSTRUCTED WITHOUT COMPLIANCE WITH FEDERAL ENVIRONMENTAL ACTS, THE STATE MUST REIMBURSE THE FEDERAL GOVERNMENT FOR SUMS INITIALLY ALLOCATED TO THE CENTER, BUT SUBSEQUENTLY DIVERTED TO OTHER STATE PROJECTS". THEREFORE, THE COURT REINSTATED THE COMPLAINT, AND REMANDED THE CASE FOR FURTHER PROCEEDINGS.
REFERENCE	ENVIRONMENTAL LAW REPORTER, VOL. 4, P 20369

CASE	FIGARSKY VS. HISTORIC DISTRICT COMMISSION, NO. 7894
LOCATION	NORWICH, CONNECTICUT'
DATE & STATUS	ENJOINED JUNE 15, 1976
PROJECT TYPE	HOUSE DESTRUCTION
ISSUE(S)	NHPA REQUIREMENTS, DEMOLITION OF BUILDING
SCOPE	THE COURT RULED THAT THE HISTORIC DISTRICT COMMISSION DOES NOT DEPRIVE THE PLAINTIFFS OF PROPERTY WITHOUT JUST COMPENSATION BY DENYING THEIR APPLICATION TO DEMOLISH THEIR HOUSE, WHICH IS LOCATED IN NORWICH HISTORICAL DISTRICT AND SCREENS THE TOWN GREEN FROM ENCROUCHING COMMERCIAL DISTRICT. THE COMMISSION CLEARY HAS STATUTORY POWER TO REGULATE THE DEMOLITION OF BUILDINGS IN HISTORIC DISTRICTS, SINCE ANY WORK DONE ON SUCH BUILDINGS MUST BE APPROVED BY A CERTIFICATE OF APPROPRIATENESS ISSUED BY THE COMMISSION. THE PLAINTIFFS CLAIMED THAT THE NORWICH HISTORIC DISTRICT ORDINANCE IS UNCONSTITUTIONAL AS APPLIED AND DEVALUES THEIR PROPERTY. THE DEFENSE ARGUED THAT PRESERVATION OF A BUILDING WHICH IS NOT ITSELF OF GREAT HISTORICAL SIGNIFICANCE BUT WHICH IS IN A DISTRICT OF EXCEPTIONAL HISTORICAL AND ARCHITECTURAL SIGNIFICANCE SERVES THE PUBLIC WELFARE. THE COURT HELD THAT THE PLAINTIFFS FAILED TO SHOW THAT THEIR HOUSE WOULD BE WITHOUT VALUE IF THEY MADE THE REPAIRS ORDERED BY THE COMMISSION. THUS, THE COURT DISMISSED THE OPPONENT'S PETITION, PROHIBITING HOUSE DESTRUCTION.
REFERENCE	ENVIRONMENTAL LAW REPORTER, VOL. 6, P 20654

CASE	FORT STORY - ITS FUTURE VS. SCHLESINGER, NO. 74-232-N
LOCATION	FORT STORY, VIRGINIA
DATE & STATUS	ENJOINED AUGUST 7, 1974
PROJECT TYPE	600 UNIT HOUSING PROJECT
ISSUE(S)	NEPA REQUIREMENTS, HIGH DENSITY DEVELOPMENT
SCOPE	RESIDENTS OF A NEARBY CITY ASSERTED THAT SINCE THE PROJECT AND ITS RESIDENTS WILL ADVERSELY AFFECT THE ECOSYSTEM OF AN ADJOINING STATE SEASHORE PARK WHICH PLAINTIFFS USE FOR RECREATION, AND WILL PLACE GREATER DEMANDS ON THE ALREADY OVER-BURDENED MUNICIPAL SERVICES AND FACILITIES IN PLAINTIFFS' CITY, A NEPA IMPACT STATEMENT IS NECESSARY TO BE FILED. THE NAVY PROJECT IS A PLAN TO CONSTRUCT 600 ROW-TYPE, MULTI-FAMILY, FOUR AND FIVE BEDROOM HOUSING UNITS ON ABOUT 81 ACRES OF LAND. JUDGE HOFFMAN, WRITING FOR THE COURT, DECLARED THAT THE NAVY'S DETERMINATION THAT AN IMPACT STATEMENT IS NOT REQUIRED IS ARBITRARY AND UNREASONABLE. THE COURT ISSUED AN INJUNCTION PROHIBITING THE U.S. NAVY FROM AWARDING CONTRACTS UNTIL AN EIS HAD BEEN FILED.
REFERENCE	ENVIRONMENTAL LAW REPORTER, VOL. 4, PP 20820-20822

CASE	FRIENDSHIP HEIGHTS CITIZENS COORDINATING COMMITTEE VS. ROZANSKY & KAY CONSTRUCTION COMPANY
LOCATION	FRIENDSHIP HEIGHTS, MD.
DATE & STATUS	ENJOINED ON 24 SEPT. 1973
PROJECT TYPE	HIGH-RISE CONDOMINIUM APARTMENT AND GARAGE
ISSUE(S)	AIR POLLUTION, SEWAGE DISPOSAL
SCOPE	CITING THE FAILURE TO CONSIDER EFFECTS ON AIR POLLUTION AND SEWAGE DISPOSAL, THE MONTGOMERY COUNTY, MD. BOARD OF APPEALS REVERSED AN ORDER ISSUING A BUILDING PERMIT TO A 17-STORY CONDOMINIUM APARTMENT BUILDING THAT IS ALREADY WELL UNDER CONSTRUCTION: THE BUILDING'S SUPERSTRUCTURE HAS BEEN COMPLETED. A CITIZENS GROUP, THE FRIENDSHIP HEIGHTS CITIZENS COORDINATING COMMITTEE, APPEALED THE COUNTY DEPARTMENT OF ENVIRONMENTAL PROTECTION'S DECISION TO GRANT THE PROJECT A BUILDING PERMIT. THE APPEALS BOARD NOTED THAT STATE REGULATIONS CLASSIFY AIR POLLUTION IN THE FRIENDSHIP HEIGHTS AREA AS "SERIOUS," AND REQUIRE THAT POLLUTION BE REDUCED AS FAST AS POSSIBLE. THE BOARD STATED THAT IT "FAILS TO PERCEIVE" HOW CONSTRUCTION OF THE PROJECT'S PLANNED 676-CAR GARAGE WILL CONTRIBUTE TO MEETING CLEAN AIR GOALS. FAILURE TO CONSIDER THE IMPACT OF THE GARAGE AND ADDITIONAL CARS BEFORE ISSUING THE PERMIT RENDERS THE PERMIT DEFECTIVE, THE BOARD HELD . THE BOARD ALSO CITED THE ADDITIONAL SEWAGE THAT WOULD BE GENERATED BY THE BUILDING, NOTING THAT THE COUNTY'S SANITARY TREATMENT COMMISSION CURRENTLY IS EXCEEDING ITS ALLOTTED TREATMENT CAPACITY AT THE AREA'S TREATMENT PLANT. ALTHOUGH THE DEVELOPERS OF THE PROJECT, ROZANSKY & KAY CONSTRUCTION COMPANY, RECEIVED PERMISSION FOR A SEWER HOOKUP, THE BOARD SAID THAT THE STANDARDS APPLIED BY THE SANITARY COMMISSION IN GRANTING SUCH "HARDSHIP" EXEMPTIONS "ARE PURELY FINANCIAL AND FAIL TO CONSIDER MATTERS OF PUBLIC HEALTH AND SAFETY." THE BOARD SAID THE ISSUE IN THE CASE IS "GRASS ROOTS RESPONSIBILITY FOR A CLEAN AND HEALTHFUL ENVIRONMENT," AND IT STATED THAT "WITHHOLDING A BUILDING PERMIT WHEN ITS ISSUANCE WOULD VIOLATE LAWS APPLYING TO POLLUTION STANDARDS IS ONE OF THE KEYS TO SURVIVAL."
REFERENCE	ENVIRONMENT REPORTER, PUBLISHED BY THE BUREAU OF NATIONAL AFFAIRS, INC., VOL. 4, NO. 22, SEPT. 28, 1973, PP 875-876

CASE	GETTYSBURG TOWER
LOCATION	GETTYSBURG, PENNSYLVANIA
DATE & STATUS	FIRST APPROVED ON 8 JUNE, 1973
PROJECT TYPE	OBSERVATION TOWER
ISSUE(S)	ENVIRONMENTAL RIGHTS AMENDMENT TO THE PENNSYLVANIA CONSTITUTION, LAND EXCHANGE
SCOPE	FINDING THAT A 1971 ENVIRONMENTAL RIGHTS AMENDMENT TO THE PENNSYLVANIA CONSTITUTION WAS NOT SELF-EXECUTING, THE PENNSYLVANIA SUPREME COURT APPROVED CONSTRUCTION OF A COMMERCIAL OBSERVATION TOWER IN GETTYSBURG. THE COURT, ALTHOUGH AFFIRMING THE DECISION OF THE PENNSYLVANIA COMMONWEALTH COURT APPROVING CONSTRUCTION OF THE TOWER, REVERSED THE FINDING OF THE PENNSYSVANIA COURT OF COMMON PLEAS THAT THE AMENDMENT WAS ENFORCEABLE WITHOUT ENABLING LEGISLATION. IN AN OPINION BY JUDGE HENRY X. O'BRIEN, THE COURT RULED THAT THE AMENDMENT WHICH GRANTS A RIGHT TO THE "PRESERVATION OF THE NATURAL, SCENIC, HISTORIC, AND AESTHETIC VALUES OF THE ENVIRONMENT," IS VAGUE AND VIOLATES THE DUE PROCESS AND EQUAL PROTECTION CLAUSES OF THE U.S. CONSTITUTION. THE AMENDMENT, O'BRIEN FOUND, FAILS TO DEFINE THE VALUES IT SEEKS TO PROTECT AND LACKS PROCEDURES BY WHICH THE USE OF PRIVATE PROPERTY CAN BE REGULATED TO PROTECT THOSE VALUES. PENNSYLVANIA ATTORNEY GENERAL ISRAEL PACKEL SAID THE STATE WOULD CONTINUE IN ITS BATTLE AGAINST THE TOWER IN FEDERAL COURT, CHARGING THAT A LAND EXCHANGE BETWEEN THE NATIONAL PARK SERVICE AND THOMAS OTTENSTEIN, BUILDER OF THE TOWER, VIOLATED THE NEPA.
REFERENCE	ENVIRONMENT REPORTER, PUBLISHED BY THE BUREAU OF NATIONAL AFFAIRS, INC., VOL. 4, NO. 24, OCT. 27, 1973, P 924 FOR THE FULL TEXT OF THE COURT'S OPINION, SEE: ENVIRONMENT REPORTER CASES, VOL. 3, P 1270

CASE	GROTON, CONNECTICUT VS. NAVY
LOCATION	GROTON, CONN.
DATE & STATUS	REINSTATED ON 18 OCTOBER, 1972
PROJECT TYPE	NAVY HOUSING PROJECT
ISSUE(S)	NEPA APPLICABLIITY
SCOPE	A FEDERAL DISTRICT JUDGE IN HARTFORD, CONNECTICUT REFUSED TO ISSUE A TEMPORARY RESTRAINING ORDER THAT WOULD BAR FURTHER CONSTRUCTION OF A 300 UNIT NAVY HOUSING PROJECT. SUIT TO BLOCK THE BAILEY HILL HOUSING PROJECT WAS FILED BY GROTON OFFICIALS, WHO CHARGED THAT THE NAVY IS REQUIRED TO FILE A NEPA IMPACT STATEMENT PRIOR TO CONSTRUCTING THE PROJECT. NAVY OFFICIALS CONTENDED THAT AN EIS WAS NOT NECESSARY BECAUSE THE PROJECT POSED NO THREAT TO THE ENVIRONMENT. WHILE RULING AGAINST THE PLAINTIFFS, CHIEF JUDGE M. JOSEPH BLUMENFELD TOOK UNDER ADVISEMENT A MOTION FOR A PERMANENT INJUNCTION THAT WOULD BLOCK CONSTRUCTION COMPLETELY.
REFERENCE	ENVIRONMENT REPORTER, PUBLISHED BY THE BUREAU OF NATIONAL AFFAIRS, INC., VOL. 3, NO. 27, NOV. 3, 1972, P. 770

CASE	HIRAM CLARKE CIVIC CLUB, INC. VS. LYNN, NO. 72-1268
LOCATION	HOUSTON, TEXAS
DATE & STATUS	REINSTATED ON APRIL 3, 1973
PROJECT TYPE	APARTMENT COMPLEX
ISSUE(S)	NEPA APPLICABILITY

SCOPE

THE U.S. COURT OF APPEALS FOR THE FIFTH CIRCUIT UPHELD A LOWER COURT DECISION THAT HUD'S THRESHOLD DETERMINATION NOT TO FILE AN IMPACT STATEMENT UNDER NEPA FOR A PROPOSED HOUSING PROJECT WAS REASONABLE.

APPELLANTS, HOMEOWNERS IN THE IMMEDIATE AREA SURROUNDING THE PROPOSED SITE, SOUGHT TO ENJOIN FEDERAL FUNDING OF THE PROJECT, ARGUING THAT HUD'S FAILURE TO FILE AN IMPACT STATEMENT BARS IT FROM FURTHER FUNDING OF THE PROJECT.

THE PROJECT IS A 272-UNIT APARTMENT COMPLEX FOR PEOPLE WITH LOW TO MODERATE INCOMES. IN ADDITION TO APARTMENT UNITS, THE COMPLEX IS TO CONTAIN COURTYARDS, OPEN SPACES, CHILDREN'S PLAYGROUNDS, A SWIMMING POOL, AND A COMMUNITY CENTER.

HUD HAD ORIGINALLY PREPARED AN "ENVIRONMENTAL CLEARANCE WORK-SHEET" WHICH CONSIDERED POPULATION DENSITY AND DISTRIBUTION, ADEQUACY OF SEWER AND WATER FACILITIES, VEHICULAR TRAFFIC, INGRESS AND EGRESS, NOXIOUS ODORS, INHARMONIOUS PROPERTY USES, DETERIORATING NEIGHBORHOOD INFLUENCES, ADEQUACY OF COMMUNITY SUPPORT FACILITIES, LOCATION IN RELATION TO THE GENERAL AREA, THE PROPOSED MANAGEMENT PLAN FOR MAINTAINING THE APARTMENTS, AND THE VALUE OF THE PROPERTY AND THE KIND OF DEVELOPMENT THAT COULD OTHERWISE BE EXPECTED ON THE PROPERTY. ON THE BASIS OF THIS "WORKSHEET", IT WAS CONCLUDED THAT THE PROJECT WOULD NOT HAVE A SIGNIFICANT ENVIRONMENTAL IMPACT AND A "NEGATIVE IMPACT" STATEMENT WAS PREPARED AND ENDORSED BY HUD.

IN RULING THAT HUD'S ACTION WAS NOT UNREASONABLE, JUDGE RONEY NOTED THAT "APPELLANTS HAVE RAISED NO ENVIRONMENTAL FACTORS, EITHER BENEFICIAL OR ADVERSE, THAT WERE NOT CONSIDERED BY HUD..."

APPELLANTS ARGUED THAT HUD DID NOT COMPLY WITH THE FINAL COUNCIL ON ENVIRONMENTAL QUALITY (CEQ) GUIDELINES WHICH REQUIRE AN EIS TO BE PREPARED WHEN A PROPOSED FEDERAL ACTION "IS LIKELY TO BE HIGHLY CONTROVERSIAL". THE COURT SAID, HOWEVER, THAT THE CEQ GUIDELINES ARE MERELY ADVISORY.

APPELLANTS CONTENTION THAT HUD INCORRECTLY EQUATES "SIGNIFICANT EFFECT" WITH "ADVERSE EFFECT" AND THAT HUD'S INVESTIGATION WAS NECESSARILY INCOMPLETE BECAUSE THE CONCERN WAS ONLY WITH ADVERSE IMPACT LED THE COURT TO SAY THAT IT "RAISES SERIOUS QUESTIONS ABOUT THE ADEQUACY OF THE INVESTIGATORY BASIS UNDERLYING THE HUD DECISION NOT TO FILE AN ENVIRONMENTAL IMPACT STATEMENT. ...CONGRESS WAS NOT ONLY CONCERNED WITH JUST ADVERSE EFFECTS BUT WITH ALL POTENTIAL ENVIRONMENTAL EFFECTS THAT AFFECT THE QUALITY OF THE HUMAN ENVIRONMENT". NEVERTHELESS, THE COURT RULED, THIS DOES NOT REQUIRE THE PREPARATION OF AN EIS FOR THIS PROJECT.

REFERENCE ENVIRONMENTAL LAW REPORTER, VOL. 3, PP. 20287-20290

CASE	IRISH VS. GREEN, NO. 162-3
LOCATION	LITTLE TRAVERSE BAY, MICHIGAN
DATE & STATUS	REINSTATED JULY 15, 1972
PROJECT TYPE	HOUSING DEVELOPMENT
ISSUE(S)	AESTHETICS, NEPA REQUIREMENTS, WATER POLLUTION, LAND USE, TRAFFIC
SCOPE	PLAINTIFFS CHALLENGED THE PROPOSED HOUSING DEVELOPMENT ON LITTLE TRAVERSE BAY. THE COMPLAINT ARGUED THAT SEEPAGE FROM SEPTIC TANKS WAS LIKELY TO CAUSE POLLUTION OF BOTH GROUND WATER AND THE BAY; THE PROJECT WOULD RESULT IN TRAFFIC CONGESTION ON A SCENIC ACCESS ROAD; THE DEFENDANTS HAD FAILED TO PREPARE AN EIS; AND THE DEVELOPMENT WAS A WASTEFUL USE OF LAND AND SHOULD BE MODIFIED TO ALLOW FOR CLUSTER DEVELOPMENT. THE JUDGE OVERRULED THE DEFENDANTS' OBJECTIONS TO EVIDENCE OF THE BROAD POTENTIAL IMPACTS OF THE PROJECT. THE COURT'S TWELVE PAGE OPINION AGREED WITH THE PLAINTIFFS' CLAIMS ON THE WATER POLLUTION AND ACCESS ROAD ISSUES, BUT DISMISSED THEIR CLAIMS SEEKING CLUSTER DEVELOPMENT AND EIS. THE COURT ORDERED THAT CONSTRUCTION COULD PROCEED ON NO MORE THAN 40% OF THE LOTS UNTIL CENTRAL WATER AND SEWER FACILITIES WERE PROVIDED, REQUIRED THE OPENING OF ANOTHER ENTRANCE ROAD TO LESSEN TRAFFIC BURDEN ON THE SCENIC HIGHWAY, AND IMPOSED RESTRICTIONS TO LESSEN THE FLOW OF SURFACE WATER WITHIN THE PROPOSED SUBDIVISION.
REFERENCE	ECOLOGY LAW QUARTERLY, 4:1, 1974, PP. 29-30

CASE	JAIL CONSTRUCTION
LOCATION	CHICAGO, ILL.
DATE & STATUS	REINSTATED ON 13 SEPTEMBER, 1973
PROJECT TYPE	FEDERAL DETENTION CENTER AND PARKING FACILITY
ISSUE(S)	IMPACT STATEMENT REQUIREMENT, LOCATION IN RESIDENTIAL AREA
SCOPE	THE U.S. COURT OF APPEALS FOR THE SEVENTH CIRCUIT RULED THAT THE GENERAL SERVICES ADMINISTRATION DID NOT ACT ARBITRARILY AND CAPRICIOUSLY IN DETERMINING THAT THE CONSTRUCTION OF A FEDERAL DETENTION CENTER AND PARKING FACILITY IN CHICAGO, ILL. WOULD NOT REQUIRE PREPARATION OF AN ENVIRONMENTAL IMPACT STATEMENT, UPHOLDING AN EARLIER RULING, BY JUDGE FRANK MCGARR OF THE U.S. DISTRICT COURT FOR THE NORTHERN DISTRICT OF ILLINOIS. THE APPEALS COURT COMPARED THE PROPOSED CHICAGO CENTER WITH THE PROPOSED FOLEY SQUARE DETENTION CENTER TO BE BUILT IN NEW YORK CITY THAT WAS THE SUBJECT OF MUCH LITIGATION IN HANLY V. KLEINDIENST. THE COURT NOTED THAT THE MAJOR ENVIRONMENTAL OBSTACLE TO CONSTRUCTION OF THE NEW YORK CENTER WAS ITS LOCATION IN A RESIDENTIAL AREA, WHILE THE CHICAGO COMPLEX WOULD NOT BE LOCATED NEAR RESIDENTIAL DWELLINGS, BUT WOULD IN FACT SERVE TO IMPROVE A BLIGHTED AREA OF THE CITY.
REFERENCE	ENVIRONMENT REPORTER, PUBLISHED BY THE BUREAU OF NATIONAL AFFAIRS, INC., VOL. 4, NO. 21, SEPT. 21, 1973, P. 827 FOR THE FULL TEXT OF THE COURT'S OPINION IN THE PRECIDENT CASE, HANLY V. KLEINDIENST, SEE: ENVIRONMENT REPORTER CASES, VOL. 4, P. 1152

CASE	JAMES VS. LYNN, NO. 74-M-309
LOCATION	DENVER, COLORADO
DATE & STATUS	REINSTATED MAY 7, 1974
PROJECT TYPE	DEMOLITION OF HISTORICAL STRUCTURES FOR URBAN RENEWAL
ISSUE(S)	ADEQUACY OF EIS, HISTORICAL SIGNIFICANCE
SCOPE	THE COURT DENIED A MOTION BY PLAINTIFFS WITHIN A PROPOSED URBAN RENEWAL AREA FOR A PRELIMINARY INJUNCTION AGAINST FURTHER DEMOLITION ON THE GROUND THAT HUD'S EIS FOR THE PROJECT IS INADEQUATE UNDER NEPA. THE COURT RULED THAT THE EIS SUFFICIENTLY DISCUSSES EACH OF THE FIVE MATTERS WHICH MUST BE CONSIDERED ACCORDING TO THE STATUTE, AND THAT ITS ONLY INADEQUACY IS ITS FAILURE TO CONSIDER THE POSSIBILITY THAT STRUCTURES OF HISTORIC OR ARCHITECTURAL VALUE MIGHT BE DESTROYED DURING THE PROCESS OF URBAN RENEWAL. THE DEFENDANTS, HOWEVER, HAD NOT YET DESTROYED THOSE STRUCTURES HAVING HISTORIC SIGNIFICANCE. HUD HAS TIME TO MODIFY THE PROJECT ON THE BASIS OF THE INFORMATION GAINED FROM THE HEARING, THE COURT DECLARED.
REFERENCE	ENVIRONMENTAL LAW REPORTER, VOL. 4, PP 20675-6

CASE	JONES VS. LYNN, NO. 73-1057
LOCATION	BOSTON, MASSACHUSETTS
DATE & STATUS	ENJOINED ON MARCH 22, 1973
PROJECT TYPE	URBAN RENEWAL
ISSUE(S)	NEPA APPLICABILITY, DEMOLITION OF BUILDINGS

SCOPE

THE U.S. COURT OF APPEALS FOR THE FIRST CIRCUIT RULED THAT NEPA APPLIES TO AN ONGOING URBAN RENEWAL PLAN APPROVED BY HUD BEFORE THE EFFECTIVE DATE OF THE ACT IF THERE HAS BEEN MAJOR FEDERAL ACTION SINCE 1970, OR IF HUD CAN SIGNIFICANTLY ALTER THE COURSE OR TIMING OF THE DEVELOPMENT, OR IF THERE IS A REASONABLE EXPECTATION OF FURTHER MAJOR FEDERAL ACTION. THE CASE WOULD BE REMANDED TO THE DISTRICT COURT FOR DETERMINATION IF ONE OF THESE CONDITIONS APPLIES IN THIS CASE.

APPELLANTS ARE RESIDENTS OF THE AREA OF THE FENWAY URBAN RENEWAL PLAN. APPELLEES ARE HUD, THE BOSTON REDEVELOPMENT AUTHORITY (BRA), AND THE FIRST CHURCH OF CHRIST, SCIENTIST, A MAJOR OCCUPANT OF THE AREA AND SOURCE OF MUCH OF THE PLANNING.

WHILE APPELLANTS SOUGHT TO HALT FURTHER DEMOLITION AND CONSTRUCTION UNTIL AN ENVIRONMENTAL IMPACT STATEMENT WAS PREPARED, HUD ARGUED THAT SINCE THE URBAN RENEWAL PLAN WAS APPROVED IN 1967, IT WAS EXEMPT FROM THE REQUIREMENTS OF NEPA.

IN HIS OPINION, CHIEF JUSTICE COFFIN DISMISSED APPELLEE'S CONTENTION OUT-OF-HAND, SAYING, "AN ENVIRONMENTAL DISTURBANCE IS NO LESS CONSEQUENTIAL BECAUSE OF THE DATE OF ITS PLANNING".

THE COURT NOTED THAT "HUD HAS RECENTLY APPROVED THE APPLICATIONS FOR FEDERAL MORTGAGE INSURANCE FOR TWO BUILDINGS AND HAS STIPULATED TO THE PREPARATION OF A MODIFIED IMPACT STATEMENT AS TO THESE TWO PARCELS. IT IS MAINTAINED THAT APPLICATIONS FOR FEDERAL MORTGAGE INSURANCE ARE PENDING FOR TWO OTHER PLANNED EDIFICES, AND AT ORAL ARGUMENT BRA REPRESENTED THAT VIRTUALLY EVERY NEW BUILDING WILL BE THE SUBJECT OF SIMILAR APPLICATIONS. IT DOES NOT APPEAR THAT THE DISTRICT COURT CONSIDERED THE RELEVANCE OF THESE CONTENTIONS WHICH, IF TRUE, WOULD SEEM TO CALL FOR A COMPREHENSIVE STUDY OF THE UNCOMPLETED ASPECTS OF THE ENTIRE PROJECT AS BEING MOST CONSISTENT WITH THE OBJECT OF EVALUATING THE 'CUMULATIVELY SIGNIFICANT IMPACT' OF THE ANTICIPATED FEDERAL ROLE".

THE COURT GAVE HUD 90 DAYS TO COMPLETE A DRAFT ENVIRONMENTAL STATEMENT, WHILE IMMEDIATELY ENJOINING ALL WORK ON THE SYMPHONY AREA OF THE FENWAY PROJECT.

REFERENCE

ENVIRONMENTAL LAW REPORTER, VOL. 3, PP 20358-20362

FOR THE DISTRICT COURT DECISION, SEE: ENVIRONMENTAL LAW REPORTER, VOL. 3, P 20353

CASE	KENT COUNTY COUNCIL FOR HISTORIC PRESERVATION VS. ROMNEY, NO. 304 F. SUPP. 885
LOCATION	MICHIGAN
DATE & STATUS	REINSTATED 1969
PROJECT TYPE	URBAN RENEWAL
ISSUE(S)	NHPA REQUIREMENTS, CULTURAL ENVIRONMENT PRESERVATION
SCOPE	UNDER SECTION 106 OF THE NATIONAL HISTORICAL PRESERVATION ACT OF 1966, THE SUBSTANCE OF THE PROTECTIVE PROVISIONS IS DELINEATED: THE HEAD OF ANY FEDERAL AGENCY HAVING DIRECT OR INDIRECT JURISDICTION OVER A PROPOSED FEDERAL OR FEDERALLY ASSISTED UNDERTAKING IN ANY STATE, AND THE HEAD OF A FEDERAL DEPARTMENT OR INDEPENDENT AGENCY, HAVING AUTHORITY TO LICENSE ANY UNDERTAKING, SHALL, PRIOR TO THE APPROVAL OF THE EXPENDITURE OF ANY FEDERAL FUNDS ...., TAKE INTO ACCOUNT THE EFFECT OF THE PROJECT ON ANY DISTRICT, SITE, BUILDING, ... THAT IS INCLUDED IN THE NATIONAL REGISTER, AND INFORM THE ADVISORY COUNCIL ON HISTORIC PRESERVATION FOR COMMENTS. CITIZENS CHALLENGED THE FAILURE OF THE DEPARTMENT OF HOUSING AND URBAN DEVELOPMENT (HUD) TO COMPLY WITH SECTION 106 IN AN URBAN RENEWAL PROJECT WHERE PROPERTIES HAD NOT BEEN LISTED ON THE NATIONAL REGISTER UNTIL AFTER HUD'S EXECUTION OF THE LOAN AND GRANT CONTRACT. THE COURT, REJECTING THE PLAINTIFFS' CONTENTION THAT ANY INSTALLMENT OR INDIVIDUAL PAYMENT BY HUD PURSUANT TO THE CONTRACT IS AN "APPROVAL" AND THEREFORE SUBJECT TO SECTION 106 REVIEW, STATED THAT AS HUD'S EXECUTION OF THE LOAN AND GRANT CONTRACT CONSTITUTED THE AGENCY APPROVAL AND AS THAT APPROVAL PREDATED LISTING OF THE AFFECTED PROPERTIES ON THE NATIONAL REGISTER, HUD WAS UNDER NO DUTY TO COMPLY WITH SECTION 106. THUS, THE COURT FOUND THAT HUD'S EXECUTION OF A MINISTERIAL OR CLERICAL ACTION WAS NOT THE KIND OF DISCRETIONARY AUTHORITY ENVISIONED UNDER THE NATIONAL HISTORICAL PRESERVATION ACT OF 1966 AND THEREFORE NOT SUBJECT TO SECTION 106 REVIEW.
REFERENCE	WAKE FOREST LAW REVIEW, VOL. 12, SPRING 1976, PP. 51-52

CASE	LEAGUE TO SAVE LAKE TAHOE VS. TROUNDAY
LOCATION	LAKE TAHOE, NEVADA
DATE & STATUS	SUIT FILED MAY 3, 1976
PROJECT TYPE	HOTEL AND CASINO CONSTRUCTION
ISSUE(S)	VIOLATION OF STATE AIR LAWS, CLEAN AIR ACT

SCOPE

A SUIT TO BLOCK CONSTRUCTION OF TWO LAKE TAHOE, NEVADA HOTELS AND CASINOS, WHICH ALLEGEDLY WILL CAUSE VIOLATIONS OF CLEAN AIR ACT AMBIENT AIR QUALITY STANDARDS, WAS FILED MAY 3, 1976 IN THE U.S. DISTRICT COURT FOR THE DISTRICT OF NEVADA.

THE SUIT WAS FILED UNDER SECTION 304 OF THE ACT BY THE LEAGUE TO SAVE LAKE TAHOE AND SEVERAL RESIDENTS OF THE TAHOE BASIN. THEY SEEK TO HALT CONSTRUCTION OF THE 960-ROOM HOTEL OLIVER AND THE 560-ROOM TAHOE PALACE.

THE SUIT CHARGES THAT THE NEVADA DEPARTMENT OF HUMAN RESOURCES ACTED ARBITRARILY AND CAPRICIOUSLY IN GRANTING "COMPLEX SOURCE" PERMITS TO TWO HOTELS AS REQUIRED BY NEVADA REGULATIONS. THE DEPARTMENT, IN FINDING THAT THE HOTELS WOULD NOT CAUSE A VIOLATION OF THE NATIONAL AMBIENT AIR QUALITY STANDARDS, FAILED TO CONSIDER ADVERSE METEOROLOGICAL CONDITIONS AND FAILED TO CONSIDER CARBON MONOXIDE LEVELS WITHIN THE PROJECT AREAS, ACCORDING TO THE LAWSUIT.

THE DEPARTMENT, ACCORDING TO THE SUIT, HAS "IGNORED EVIDENCE" THAT THE CARBON MONOXIDE STANDARDS ARE ALREADY BEING EXCEEDED IN THE AREA, AND THAT "THE ADDITIONAL TRAFFIC ON THE HIGHWAY GENERATED BY THE TWO CASINOS WOULD CAUSE EVEN MORE SEVERE AND FREQUENT VIOLATIONS".

REFERENCE

ENVIRONMENT REPORTER, PUBLISHED BY THE BUREAU OF NATIONAL AFFAIRS, INC., VOL. 7, NO. 2, MAY 14, 1976, P. 31

CASE	LITTLE WOLF LAKE PROPERTY OWNERS ASSOCIATION VS. HAASE, NO. 74-837-CE
LOCATION	MICHIGAN
DATE & STATUS	ENJOINED JULY 25, 1975
PROJECT TYPE	78-LOT REAL ESTATE DEVELOPMENT CONSTRUCTION
ISSUE(S)	ZONING CODES, NEPA REQUIREMENTS
SCOPE	THE PROJECT, WHICH WOULD "FUNNEL ACCESS" TO A LAKE THROUGH ONLY NINE RIPARIAN LOTS, WAS CHALLENGED BY PLAINTIFFS BECAUSE OF ITS SIZE. THE PROPOSED DEVELOPMENT, THEY CLAIMED, WAS SUBJECT TO THE SUBDIVISIONS CONTROL ACT OF 1967 WHICH REQUIRES TOWNSHIP APPROVAL OF REGULATED PROJECTS. THE TOWNSHIP DEFENDANT MOVED FOR A SUMMARY JUDGEMENT ON THE GROUND THAT IT EXERCISES NO STATUTORY DUTIES UNDER MEPA TO CONSIDER ENVIRONMENTAL MATTERS WHEN APPROVING PLOTS UNDER THE SUBDIVISIONS CONTROL ACT. THE COURT DENIED THE MOTION AND ENJOINED THE CONSTRUCTION UNTIL MEPA CONSIDERATIONS WERE MADE BY THE DEFENDANTS.
REFERENCE	JOURNAL OF URBAN LAW, VOL. 53:589, 1976, P. 615

CASE MAHER VS. CITY OF NEW ORLEANS, NO. 516 F. 2D 1051

LOCATION NEW ORLEANS

DATE & STATUS REJECTED 1975

PROJECT TYPE RENT-PRODUCING APARTMENTS

ISSUE(S) NHPA, ZONING CODES, ALTERNATIVE CONSIDERATIONS

SCOPE BECAUSE OF THE HISTORICAL SIGNIFICANCE OF THE VIEUX CARRE SECTION OF NEW ORLEANS, THE CITY PASSED AN HISTORICAL ZONING ORDINANCE. THE PLAINTIFF ARGUED THAT THE VIEUX CARRE ORDINANCE MADE HIS PROPERTY PRACTICALLY UNUSABLE. THUS, THE ORDINANCE AMOUNTED TO CONFISCATION, IN HIS OPINION.

THE PLAINTIFF HAS A COTTAGE AND RESIDENCE ON SOME PROPERTY IN THE VIEUX CARRE SECTION WHICH HE WISHED TO DEMOLISH AND REPLACE WITH RENT-PRODUCING APARTMENTS.

THE COURT, DECIDING THAT THE ORDINANCE WAS NOT CONFISCATORY, DENIED PERMISSION FOR DEMOLITION. THE COURT FOUND THAT THE PLAINTIFF HAD NOT MADE ANY EFFORT TO RENT THE COTTAGE FOR AS MUCH BENEFIT AS HE COULD HAVE, AND CONCLUDED THAT THE PLAINTIFF HAD WHOLLY FAILED TO CARRY HIS BURDEN OF PROVING HE HAD BEEN DEPRIVED OF ANY REASONABLE USE OR RETURN FROM HIS PROPERTY.

REFERENCE WAKE FOREST LAW REVIEW, VOL. 12, SPRING 1976, PP. 243-244

CASE	MARYLAND PLANNING COMMISSION VS. U.S. POSTAL SERVICE
LOCATION	PRINCE GEORGES COUNTY, MD.
DATE & STATUS	FILED 1973
PROJECT TYPE	BULK MAIL FACILITY
ISSUE(S)	ZONING ORDINANCE, LAND USE REGULATION

SCOPE

CONFORMITY TO OR DEVIATION FROM LOCAL LAND USE REGULATIONS MUST BE CONSIDERED IN DETERMINING WHETHER A PROJECT "SIGNIFICANTLY" AFFECTS THE QUALITY OF THE HUMAN ENVIRONMENT, ACCORDING TO AN AUGUST 23 DECISION OF THE U.S. COURT OF APPEALS FOR THE DISTRICT OF COLUMBIA.

THE COURT RULED, CONCERNING CONSTRUCTION OF A BULK MAIL FACILITY IN PRINCE GEORGES COUNTY, MD. THE U.S. DISTRICT COURT FOR THE DISTRICT OF COLUMBIA IN OCTOBER 1972 DENIED A REQUEST FOR A TEMPORARY INJUNCTION TO HALT CONSTRUCTION OF THE FACILITY, HOLDING THAT THE POSTAL SERVICE'S FINDING THAT THE FACILITY WOULD NOT HAVE SIGNIFICANT ENVIRONMENTAL IMPACT WAS NOT ARBITRARY AND CAPRICIOUS.

THE APPEALS COURT NOTED THAT THE LOCAL PLANNING AGENCY, THE MARYLAND-NATIONAL CAPITAL PARK AND PLANNING COMMISSION, FOUND THAT THE FACILITY WOULD NOT BE WHOLLY IN CONFORMANCE WITH APPLICABLE ZONING ORDINANCES.

JUDGE HAROLD LEVENTHAL, WRITING FOR THE COURT, HELD THAT SUCH NONCONFORMITY WITH LOCAL ZONING REGULATIONS SHOULD HAVE BEEN CONSIDERED MORE CAREFULLY BY THE POSTAL SERVICE BEFORE ARRIVING AT A FINDING OF NONSIGNIFICANCE. LEVENTHAL NOTED THAT CONFORMITY WITH ZONING REGULATIONS MAY CREATE A PRESUMPTION THAT THE PROPOSED ACTION IS NOT SIGNIFICANT, SINCE THE ENVIRONMENTAL EFFECTS WILL BE NO GREATER THAN WHAT WAS CONTEMPLATED BY THE LOCAL PLANNING AGENCY. ON THE OTHER HAND, HE STATED, THE NEPA REQUIRES MORE CAREFUL SCRUTINY WHEN THE FEDERAL GOVERNMENT "EXERCISES ITS SOVEREIGNTY SO AS TO OVERRIDE LOCAL ZONING PROTECTIONS."

THE APPEALS COURT DID NOT REVERSE THE DISTRICT COURT'S DENIAL OF PREMIMINARY INJUNCTIVE RELIEF. IT NOTED THAT THE FACILITY WAS SUBSTANTIALLY COMPLETED AS OF MAY 1973 AND THAT A PRELIMINARY INJUNCTION NOW WOULD NOT BE HELPFUL. THE APPEALS COURT ALSO CONCLUDED THAT IT COULD NOT BE SURE THAT THE DISTRICT COURT HAD ABUSED ITS DISCRETION IN DETERMINING THAT THE PROJECT DID NOT REQUIRE AN IMPACT STATEMENT.

THE COURT REMANDED THE CASE TO THE DISTRICT COURT, HOWEVER, FOR A MORE CAREFUL SCRUTINY ON POTENTIAL PROBLEMS OF WATER AND OIL RUNOFF FROM THE FACILITY AND ON WHETHER THESE PROBLEMS REQUIRE A STATEMENT. THE COURT NOTED THAT IF IT WERE DECIDING THE CASE ON THE MERITS RATHER THAN ON APPEAL FROM A DENIAL OF INJUNCTIVE RELIEF, IT "WOULD BE HARD PUT TO AVOID A RULING THAT THE REQUISITE IMPACT STATEMENT HAD ERRONEOULLY BEEN BYPASSED." AT THE VERY LEAST, THE COURT SAID, THE POSTAL SERVICE'S FINDING THAT A STATEMENT WAS NOT REQUIRED SHOULD PROVIDE CONVINCING REASONS WHY A CONSTRUCTION PROJECT WITH "ARGUABLY" POTENTIALLY SIGNIFICANT ENVIRONMENTAL IMPACT DOES NOT REQUIRE A DETAILED IMPACT STATEMENT.

REFERENCE

ENVIRONMENT REPORTER, PUBLISHED BY THE BUREAU OF NATIONAL AFFAIRS, INC., VOL. 4, NO. 18, AUG. 31, 1973, P. 729

CASE	MCCORMICK VS. LAWRENCE
LOCATION	OSEETAH LAKE, ADIRONDAK PARK, NEW YORK
DATE & STATUS	ENJOINED ON 11 AUGUST, 1975
PROJECT TYPE	BOATHOUSES
ISSUE(S)	AESTHETICS, NATURAL BEAUTY

SCOPE

THE ADIRONDAK PARK AGENCY'S DENIAL OF PERMISSION FOR PRIVATE DEVELOPERS TO CONSTRUCT BOATHOUSES ON OSEETAH LAKE WAS DECLARED LAWFUL AND NOT ARBITRARY BY THE NEW YORK SUPREME COURT ON 11 AUGUST, 1975.

IT WAS NOTED THAT THE AGENCY HAD BEEN CREATED BY THE NEW YORK LEGISLATURE FOR THE PURPOSE OF PROVIDING OPTIMUM CONSERVATION, PROTECTION, PRESERVATION AND DEVELOPMENT OF THE PARK AREA.

JUDGE EDMUND L. SHEA SAID, "(H)OWEVER RELUCTANT COURTS HAVE BEEN IN THE PAST TO ALLOW AESTHETIC CONSIDERATIONS ALONE TO JUSTIFY THE USE OF POLICE POWER, THE COURTS NOW RECOGNIZE AESTHETICS AS A LEGITIMATE CONCEPT WITHIN THE GENERAL POLICE POWER.

"IT CAN BE SEEN THAT THE VIEW FROM THE MAIN CHANNEL (OF OSEETAH LAKE), WHICH PASSES CLOSE BY THE PETITIONER'S PROPERTY, STILL RETAINS A SCENIC VIEW OF THE TREES AND SHORELINE UNSPOILED BY OVER-DEVELOPMENT. ...THE ADDITION OF A CLUSTER OF BOATHOUSES ON A PROJECT OF THIS SIZE WOULD TEND TO DESTROY THE NATURAL BEAUTY OF THE AREA, AND THAT THE AGENCY'S DECISION CANNOT BE DEEMED ARBITRARY OR CAPRICIOUS."

IT WAS NOTED ALSO THAT ALTHOUGH BOATHOUSES ARE EXEMPTED FROM SET-BACK RESTRICTIONS BY STATE LAW, THAT DOESN'T MEAN THAT THEY CANNOT BE PROHIBITED ENTIRRLY.

REFERENCE

ENVIRONMENT REPORTER, PUBLISHED BY THE BUREAU OF NATIONAL AFFAIRS INC., VOL 6, NO. 19, SEPT. 5, 1975, PP. 763-764

CASE	MCLEAN GARDENS RESIDENTS ASSOCIATION VS. NATIONAL CAPITAL PLANNING COMMISSION
LOCATION	WASHINGTON, D.C.
DATE & STATUS	ENJOINED ON 21 OCTOBER, 1972
PROJECT TYPE	COMMERCIAL COMPLEX AND LUXURY HIGH-RISE APARTMENT COMPLEX
ISSUE(S)	NEPA APPLICABILITY, DEMOLITION OF EXISTING STRUCTURES, CONGESTION, TRAFFIC, AIR POLLUTION, SOCIAL IMPACT

SCOPE

THE DISTRICT COURT GRANTED AN INJUNCTION PROHIBITING THE RAZING OF A MEDIUM-DENSITY, LOW AND MIDDLE INCOME GARDEN APARTMENT AND DORMITORY COMPLEX TO MAKE WAY FOR A HIGH-DENSITY SHOPPING AND RESIDENTIAL COMPLEX INCLUDING MOSTLY UPPER-MIDDLE INCOME HOUSING. THE SUIT WAS BROUGHT BY THE MCLEAN GARDENS RESIDENTS ASSOCIATION, A NON-PROFIT CORPORATION REPRESENTING THE INTERESTS OF THE RESIDENTS OF MCLEAN GARDENS, AGAINST THE NATIONAL CAPITAL PLANNING COMMISSION, A FEDERAL ADMINISTRATIVE AGENCY CREATED TO PLAN THE DEVELOPMENT OF THE CAPITAL AND TO CONSERVE ITS NATURAL AND HISTORICAL FEATURES, AND THE DISTRICT OF COLUMBIA ZONING COMMISSION.

THE DEVELOPMENT PLANS CALLED FOR RAZING THE PRESENT MODERATE RENTAL UNITS AND DRASTICALLY CHANGING THE CHARACTER OF THE PRESENT GARDEN-APARTMENT PARK-LIKE SETTING TO ONE OF INTENSIVE COMMERCIAL DEVELOPMENT AND LUXURY HIGH-RISE APARTMENT UNITS. THE OWNERS PROPOSED TO BUILD A 90-FOOT OFFICE BUILDING; 350-ROOM COMMERCIAL AND RESORT HOTEL; A 2-BLOCK LONG SHOPPING MALL; AND 2,210 HIGH-RISE APARTMENT UNITS, 1,510 OF WHICH WOULD BE CONDOMINIUMS WITH SALES PRICES RANGING BETWEEN $22,500 AND $77,500 AND 700 WOULD BE RENTAL APARTMENTS WITH RENTS RANGING BETWEEN $175 AND $500 PER MONTH, NOT INCLUDING PARKING COSTS.

THE JUDGE NOTED THE FOLLOWING CONCERNS REGARDING THE IMPACT OF THE PROPOSED PROJECT ON THE HUMAN AND URBAN ENVIRONMENT: A) MCLEAN GARDENS IS BORDERED ON THE WEST BY THE GLOVER-ARCHBOLD NATIONAL PARK, RAISING QUESTIONS AS TO THE INTRUSIVEMESS OF THE DEVELOPMENT ON THE PARK, CONCERNS ALREADY RAISED BY THE NATIONAL PARK SERVICE; B) MCLEAN GARDENS PROVIDES ABOUT 2,000 UNITS OF LOW TO MODERATE INCOME HOUSING. SINCE THE CITY OF WASHINGTON IS ALREADY DEFICIENT IN ITS HOUSING SUPPLY FOR FAMILIES AND INDIVIDUALS OF LOW- OR MODERATE-INCOME LEVELS, THE WITHDRAWAL OF HOUSING UNITS WILL FURTHER INTENSIFY PRESSURES ON THOSE WITH NO HOUSING ALTERNATIVES. THE SITUATION IS FURTHER EXACERBATED IN MCLEAN GARDENS BECAUSE A SUBSTANTIAL PORTION OF ITS RESIDERTS IS OVER THE AGE OF 60; C) THE PROPOSED PROJECT WILL GENERATE SUBSTANTIAL INCREASES IN CONGESTION, TRAFFIC, AND AIR POLLUTION. THE DEVELOPER'S OWN FIGURES ESTABLISH THAT THE INCREASE IN PEAK HOUR USAGE AT A NUMBER OF MAJOR INTERSECTIONS IN THE AREA WILL DRIVE STREET CAPACITY TO OVER 100%. THE DEVELOPERS WILL BRING IN 5,000 ADDITIONAL CARS FOR WHICH PARKING WILL BE PROVIDED, THUS FURTHER INCREASING THE FLOW OF TRAFFIC AT ALL TIMES OF THE DAY AND NIGHT; D) SINCE ABOUT 98% OF ALL THE CARBON MONOXIDE IN THE DISTRICT OF COLUMBIA IS GENERATED BY AUTOMOBILES, THE INCREASE IN TRAFFIC CAN ONLY INCREASE THE DANGER TO THE HEALTH AND SAFETY OF ALL CITIZENS IN THE AFFECTED AREAS; E) SINCE INCREASED LEVELS OF AIR POLLUTION ADVERSELY EFFECT VEGETATION AND FOLIAGE, THE INCREASED POLLUTION CAUSED BY THE INCREASED TRAFFIC WOULD IMPACT THE SHRUBS, TREES AND OTHER FOLIAGE OF GLOVER-ARCHBOLD PARK; F) THE PROJECT WILL CAUSE PROBLEMS OF TRASH AND SEWAGE DISPOSAL. THE BLUE PLAINS SEWAGE TREATMENT PLANT IS ALREADY GREATLY OVERTAXED; AND G) THERE ARE QUESTIONS OF LONG RANGE PLANNING AND URBAN DEVELOPMENT.

THE COURT RULED THAT THE ROLE PLAYED BY THE NATIONAL CAPITAL PLANNING COMMISSION IN THE PROCESSING OF PLANNED UNIT DEVELOPMENT APPLICATIONS BY THE D.C. ZONING COMMISSION CONSTITUTED A SUBSTANTIAL AND "MAJOR FEDERAL ACTION", AND ENJOINED THE DEFENDENTS FROM PROCEEDING WITH ANY ASPECT OF THE PROJECT UNTIL A COMPREHENSIVE IMPACT STATEMENT IS PREPARED.

REFERENCE — ENVIRONMENTAL LAW REPORTER, VOL. 2, PP. 20659-20662

CASE	MILO COMMUNITY HOSPITAL VS. WEINBERGER, NO. 75-1205
LOCATION	MILO, MAINE
DATE & STATUS	ENJOINED NOVEMBER 14, 1975
PROJECT TYPE	HOSPITAL CERTIFICATION
ISSUE(S)	NEPA REQUIREMENTS, FIRE REGULATIONS
SCOPE	THE COURT RULED THAT THE SECRETARY OF HEALTH, EDUCATION AND WELFARE WAS NOT REQUIRED TO PREPARE A NEPA IMPACT STATEMENT PRIOR TO TERMINATING THE PLAINTIFF'S STATUS AS A FEDERALLY-ASSISTED "PROVIDER OF SERVICES" UNDER THE MEDICARE ACT BECAUSE OF THE PLAINTIFF'S CONTINUED NON-COMPLIANCE WITH FIRE PREVENTION REGULATIONS. ALTHOUGH CERTAIN DECERTIFICATION DECISIONS UNDER THE ACT MAY REQUIRE ENVIRONMENTAL IMPACT STATEMENTS UNDER NEPA, THE ACT IMPOSES ON THE SECRETARY A NON-DISCRETIONARY DUTY TO TERMINATE THE CERTIFICATION OF ANY HOSPITAL IN WHICH HE FINDS SERIOUS NON-COMPLIANCE WITH FIRE PREVENTION REQUIREMENTS. THE SECRETARY THEREFORE DID NOT ERR IN FAILING TO PREPARE AN EIS IN THIS PARTICULAR CASE BECAUSE HE WAS STATUTORILY PROHIBITED FROM TAKING INTO ACCOUNT FACTORS CHARACTERIZED BY THE PLAINTIFF AS "ENVIRONMENT CONSIDERATIONS" I.E. ECONOMIC DISLOCATIONS IN THE COMMUNITY AND INCREASED USE OF AUTOMOBILES BY RESIDENTS TRAVELING TO MORE DISTANT HOSPITALS.
REFERENCE	ENVIRONMENTAL LAW REPORTER, VOL. 6, 1975, PP. 20027-20029

CASE	MONTANA WILDERNESS ASSOCIATION VS. MONTANA BOARD OF HEALTH AND ENVIRONMENTAL SCIENCES, NO. 38902
LOCATION	GALLATIN CANYON, MONTANA
DATE & STATUS	ENJOINED AUGUST 29, 1975
PROJECT TYPE	HOUSING SUBDIVISION DEVELOPMENT
ISSUE(S)	EIS ADEQUACY, ECONOMIC FEASIBILITY, AESTHETICS, ALTERNATIVE ACTIONS, FUTURE IMPACTS
SCOPE	DEFENDANT'S IMPACT STATEMENT CONCERNING ITS APPROVAL OF A HOUSING SUBDIVISION IN GALLATIN CANYON FAILED TO MEET PROCEDURAL REQUIREMENTS OF THE MONTANA EPA, ARGUED THE MONTANA WILDERNESS ASSOCIATION. THE EIS CONTAINS NO ECONOMIC ANALYSIS, NO ATTEMPT TO QUANTIFY AESTHETIC CONSIDERATIONS SUCH AS VISUAL IMPACTS, NO SERIOUS DISCUSSION OF ALTERNATIVES, NO ANALYSIS OF COMMENTS OBJECTING TO IMPACTS ON WILDLIFE, NO EVALUATION OF TRAFFIC EFFECTS, NO DISCUSSION OF FUTURE ENERGY NEEDS, AND NO EVALUATION OF FUTURE IMPACTS. SINCE THE IMPACT STATEMENT IS PATENTLY INADEQUATE, INJUNCTIVE AS WELL AS DECLARATIVE RELIEF IS APPROPRIATE PENDING PROMULGATION OF AN ADEQUATE ENVIRONMENTAL IMPACT STATEMENT. THE EIS IS VOID; THE DEFENDANT IS ORDERED TO RESCIND ITS REMOVAL OF SANITORY CONDITIONS ON THE PLOT IN QUESTION; THE INTERVENING DEVELOPER IS ORDERED TO STOP FURTHER SUBDIVISION DEVELOPMENT. THE JUDGE, WRITING FOR THE COURT, CONSIDERED ALL THE ALLEGATIONS, AND CONCLUDED THAT "THE PROCEDURE ADOPTED IN PROMULGATING THE EIS IS WHOLLY INADEQUATE TO MEET THE STANDARDS SET BY THE STATUTE. IN VIEW OF THIS, I CONSIDER IT UNNECESSARY TO EXAMINE THE QUESTION OF WHETHER EIS CONFORMS TO THE DEFENDANT'S DEPARTMENT'S OWN REGULATION". BECAUSE THE COURT DETERMINED THAT THE STATEMENT CLEARLY SHOWS THAT IN MANY RESPECTS THE DECISION MADE WAS NOT BASED ON "INDIVIDUAL CONSIDERATION AND BALANCING OF ENVIRONMENTAL FACTORS - CONDUCTED FULLY AND IN GOOD FAITH", THE JUDGE SAW THAT AN INJUNCTION WAS REQUIRED IN ORDER TO CARRY OUT THE CLEAR, STRONG MANDATE OF THE ACT.
REFERENCE	ENVIRONMENTAL LAW REPORTER, VOL. 6, 1975, PP. 20043-20045

CASE	MUHA VS. UNION LAKES ASSOCIATION, NO. 2964
LOCATION	MICHIGAN
DATE & STATUS	FILED AUGUST, 1973
PROJECT TYPE	SHOPPING CENTER AND PARKING LOT
ISSUE(S)	WATER POLLUTION

PLAINTIFFS ARGUED THAT CONSTRUCTION INVOLVED THE DUMPING OF SILT INTO AN ADJACENT CREEK, WHICH THREATENED THE LIFE OF THE CREEK. THE CONSTRUCTION WAS ON LOW-LYING LAND NEAR A STREAM.

AFTER THE PLAINTIFFS HAD FILED THE COMPLAINT, THE DEPARTMENT OF NATURAL RESOURCES REQUESTED THE ATTORNEY GENERAL TO INITIATE ACTION TO COMPEL THE DEFENDANTS TO CLEAR THE SILT FROM THE CREEK, CUT BACK THE PARKING AREA, AND INCREASE THE LAND AREA TO BE USED FOR DRAINAGE.

REFERENCE ECOLOGY LAW QUARTERLY, VOL. 4:1, 1973, P. 12

CASE	NUCLEUS OF CHICAGO HOMEOWNERS VS. LYNN, NO. 74-1206
LOCATION	CHICAGO, ILLINOIS
DATE & STATUS	
PROJECT TYPE	LOW-INCOME PUBLIC HOUSING
ISSUE(S)	NEPA APPLICABILITY, ALTERNATIVE ACTIONS, CRIMINAL ACTS, PHYSICAL VIOLENCE, ECONOMIC DECLINE

SCOPE

THE U.S. COURT OF APPEALS UPHELD THE FEDERAL DISTRICT COURT'S DECISION THAT REFUSED TO ORDER HUD TO STUDY THE ENVIRONMENTAL IMPACT OF PROSPECTIVE PUBLIC HOUSING TENANTS.

CHICAGO COMMUNITY ORGANIZATIONS AND INDIVIDUALS SUED TO HALT THE CONSTRUCTION OF LOW-INCOME HOUSING IN VARIOUS DISTRICTS, CONTENDING THAT TENANTS OF PUBLIC HOUSING AS A "STATISTICAL WHOLE" HAVE A HIGHER PROPENSITY TOWARD CRIMINAL BEHAVIOR AND ACTS OF VIOLENCE THAN OTHER RESIDENTS, AND THAT CONSTRUCTION OF PUBLIC HOUSING WILL INCREASE THE HAZARD OF CRIMINAL ACTS, PHYSICAL VIOLENCE, AND ECONOMIC DECLINE IN SITE NEIGHBORHOODS. THE PLAINTIFFS CHARGED THAT HUD VIOLATED THE NATIONAL ENVIRONMENATL POLICY ACT (NEPA) BY PREPARING ONLY A "NEGATIVE" STATEMENT, FINDING NO ADVERSE ENVIRONMENTAL IMPACT WOULD OCCUR.

ON NOVEMBER 11, 1973, THE U.S. DISTRICT COURT FOR THE NORTHERN DIVISION OF ILLINOIS RULED AGAINST THE PLAINTIFF, FINDING THAT ALTHOUGH HUMAN BEINGS MAY BE POLLUTERS, "THEY ARE NOT THEMSELVES POLLUTION."

IN AN OPINION BY JUDGE THOMAS E. FAIRCHILD THE COURT SAID, "TO THE EXTENT THAT THIS CLAIM (THAT TENANTS OF LOW-INCOME HOUSING ARE STATISTICALLY MORE LIKELY TO COMMIT VIOLENT CRIMES) CAN BE CONSTRUED TO MEAN THAT HUD MUST CONSIDER THE FEAR OF THE NEIGHBORS OF PROSPECTIVE PUBLIC HOUSING TENANTS, WE SERIOUSLY QUESTION WHETHER SUCH AN IMPACT IS COGNIZABLE UNDER NEPA."

WHILE THE COURT DID NOT RULE ON THAT QUESTION, IT DID RULE THAT HUD DID CONSIDER, IN GOOD FAITH, THE IMPACT ON THE NEIGHBORHOOD AND DID SATISFY NEPA THAT AN IMPACT STATEMENT WAS NOT REQUIRED. THE COURT ALSO NOTED THE PREVIOUS COURT ORDER DIRECTING HUD TO BUILD SCATTERED-SITE LOW-INCOME HOUSING WHICH CURTAILED HUD'S CONSIDERATION OF ALTERNATIVES, BUT HUD DID CONSIDER ALTERNATIVE SITES.

REFERENCE

ENVIRONMENT REPORTER, PUBLISHED BY THE BUREAU OF NATIONAL AFFAIRS, INC., VOL. 6, NO. 25, OCT. 17, 1975, PP. 1012-1013

CASE	PENN CENTRAL TRANSPORTATION CO. VS. CITY OF NEW YORK, NO. 1305
LOCATION	NEW YORK CITY
DATE & STATUS	FILED DECEMBER 16, 1975
PROJECT TYPE	TOWER OFFICE BUILDING
ISSUE(S)	LANDMARK PRESERVATION, PRIVATE PROPERTY RIGHTS
SCOPE	THE COURT REVERSED A LOWER COURT DECISION UPHOLDING THE CONSTITUTIONALITY OF THE NEW YORK CITY LANDMARKS PRESERVATION LAW AS APPLIED TO GRAND CENTRAL TERMINAL, OVER WHICH PLAINTIFFS INTEND TO CONSTRUCT A 50 STORY OFFICE BUILDING. PLAINTIFFS SHOWED THAT THEY HAVE BEEN RENDERED INCAPABLE OF OBTAINING A REASONABLE RETURN ON THEIR INVESTMENT IN ORDER TO OBTAIN COMPENSATION. THE OPPOSITION ARGUED THAT THE SUBSTANTIAL OPERATING DEFICIT OF THE TERMINAL DEPRIVES THE PLAINTIFFS OF ANY REASONABLE USE. FURTHERMORE, THAT THE ORIGINAL PLANS FOR THE TERMINAL CALLED FOR A BUILDING ABOVE IT INDICATES THAT THE LANDMARKS LAW WOULD NOT BE VITIATED BY CONSTRUCTION OF THE PROPOSED OFFICE BUILDING. DESIGNATING THE TERMINAL AS A LANDMARK WOULD CONSTITUTE TAKING PLAINTIFFS' PRIVATE PROPERTY FOR PUBLIC USE WITHOUT JUST COMPENSATION
REFERENCE	ENVIRONMENTAL LAW REPORTER, VOL. 6, P 20251

CASE	PEOPLE OF SAIPAN VS. DEPARTMENT OF INTERIOR, NO. 73-1769
LOCATION	MICRO BEACH, SAIPAN
DATE & STATUS	APPROVED 16 JULY, 1974
PROJECT TYPE	HOTEL
ISSUE(S)	APPLICABILITY OF NEPA IN A U.S. TRUST TERRITORY
SCOPE	U.S. COURT OF APPEALS RULED THAT THE PACIFIC ISLANDS TRUST TERRITORY HIGH COMMISSIONER'S APPROVAL OF CONSTRUCTION OF A PROPOSED HOTEL ON MICRO BEACH, SAIPAN, IS NOT A "FEDERAL" ACTION SUBJECT TO THE REQUIREMENTS OF NEPA. THE APPEALS COURT UPHELD A MARCH, 1973 DECISION OF THE U.S. DISTRICT COURT FOR THE DISTRICT OF HAWAII WHICH RULED THAT EVEN THOUGH NEPA IS APPLICABLE TO FEDERAL AGENCIES OPERATING IN THE TRUST TERRITORY, ITS DOES NOT APPLY TO ACTIONS OF THE TRUST TERRITORY GOVERNMENT. WHILE REJECTING THE PLAINTIFF'S CLAIMS UNDER NEPA THE APPELLATE COURT, IN AN OPINION BY JUDGE ALFRED T. GOODWIN, RULED THAT THE LOCAL COURTS OF THE TRUST TERRITORY HAVE THE AUTHORITY TO DETERMINE WHETHER THE APPROVAL VIOLATED THE SUBSTANTIVE RIGHTS EMBODIED IN THE TERRITORY'S TRUSTEESHIP AGREEMENT. ALTHOUGH THESE RIGHTS IN THE TRUSTEESHIP AGREEMENT ARE NOT DEFINED PRECISELY, THE AGREEMENT DOES PROVIDE ENOUGH STANDARDS AGAINST WHICH TO TEST AN APPROVAL ALLEGEDLY MADE IN DISREGARD "OF THE PROTESTS OF THE ISLANDS' ELECTED OFFICIALS AND WITHOUT A SHOWING OF CONSIDERATION OF CULTURAL AND ENVIRONMENTAL FACTORS."
REFERENCE	ENVIRONMENT REPORTER, PUBLISHED BY THE BUREAU OF NATIONAL AFFAIRS, INC., VOL. 5 NO. 14, AUG. 2, 1974, P.430 FOR THE FULL TEXT OF THE DECISION OF THE U.S. DISTRICT COURT SEE: ENVIRONMENT REPORTER CASES, VOL. 5, P. 1159

CASE	PINKNEY VS. OHIO ENVIRONMENTAL PROTECTION AGENCY, NO. C-73-1159
LOCATION	CUYAHOGA COUNTY, OHIO
DATE & STATUS	REINSTATED JANUARY 17, 1974
PROJECT TYPE	SHOPPING MALL CONSTRUCTION
ISSUE(S)	AIR POLLUTION, TRAFFIC, ENVIRONMENTAL RIGHTS, BUILDING PERMITS

SCOPE

THE COURT DENIED THE MOTION FOR A PRELIMINARY INJUNCTION AND DISMISSED AN ACTION WHICH ALLEGED THAT THE GRANTING OF PERMITS FOR CONSTRUCTION OF A LARGE ENCLOSED SHOPPING CENTER VIOLATES THE CLEAN AIR ACT AND THE U.S. CONSTITUTION.

THE PLAINTIFFS CLAIMED THAT THE PROJECT WOULD CAUSE VIOLATION OF AMBIENT AIR QUALITY STANDARDS BY ATTRACTING AUTOMOBILES AND OVERBURDENING SEWERS, THAT OHIO HAD ILLEGALLY FAILED TO PROMULGATE REGULATIONS FOR INDIRECT SOURCE REVIEW (A PUBLIC HEARING), AND THAT THE EPA ADMINISTRATOR HAD ABUSED HIS DISCRETION IN DELAYING THE EFFECTIVE DATE OF FEDERAL INDIRECT SOURCE REVIEW RULES FOR 180 DAYS. THE PLAINTIFFS SOUGHT INJUNCTION RELIEF AGAINST GOVERNMENT OFFICIALS AND AGENCIES WHICH THEY CLAIMED HAD BEEN, WOULD BE, OR SHOULD BE INVOLVED IN THE GRANTING OF DEVELOPMENT PERMITS.

THE COURT DENIED PLAINTIFFS' MOTION IN VIEW OF PLAINTIFFS' FAILURE TO GIVE 60 DAYS NOTICE PRIOR TO FILING SUIT OR TO BRING THE ACTION IN THE U.S. COURT OF APPEALS AS IS REQUIRED BY STATUTE. JUDGE LAMBROS, WRITING FOR THE COURT, DECLARED THAT THE APA (ADMINISTRATIVE PROCEDURE ACT), WHICH PLAINTIFFS ARGUE IS APPLICABLE, DOES NOT APPLY WHERE THERE IS AN EXPLICIT PROCEDURE FOR JUDICIAL REVIEW SET FORTH FOR ADMINISTRATIVE ACTIONS TAKEN UNDER A PARTICULAR STATUTE. THE COURT RULES THAT THERE IS NO FUNDAMENTAL, CONSTITUTIONALLY PROTECTED RIGHT TO A HEALTHFUL ENVIRONMENT, AND THAT PLAINTIFFS HAVE FAILED TO ALLEGE FACTS WHICH SHOW A DEPRIVATION OF LIBERTY OR PROPERTY WITHOUT PROCEDURAL DUE PROCESS.

THUS, THE COURT RULED THAT PLAINTIFFS HAVE NOT MADE ALLEGATIONS SUFFICIENT TO DEMONSTRATE THAT THEY ARE HARMED BY THE PROCEDURES IN QUESTION, INCLUDING WITHHOLDING OF A HEARING PRIOR TO ISSUANCE OF A PERMIT.

REFERENCE: ENVIRONMENTAL LAW REPORTER, VOL. 4, P 20460

CASE	PLAN FOR ARCADIA VS. ANITA ASSOCIATES, NO. 74-1180
LOCATION	ARCADIA, CALIFORNIA
DATE & STATUS	REINSTATED ON 10 JULY, 1974
PROJECT TYPE	SHOPPING CENTER
ISSUE(S)	AIR STANDARDS, APPLICABILITY OF NEPA TO SHOPPING CENTERS
SCOPE	A FEDERAL DISTRICT COURT'S DISMISSAL OF AN ENVIRONMENTAL GROUP'S ATTEMPT UNDER THE CLEAN AIR ACT TO ENJOIN CONSTRUCTION OF A APPEALS FOR THE NINTH CIRCUIT.

A GROUP OF RESIDENTS AND PROPERTY OWNERS IN ARCADIA CALIFORNIA CHARGED THAT CONSTRUCTION OF THE SHOPPING CENTER WOULD CAUSE A VIOLATION OF THE NATIONAL AMBIENT AIR STANDARDS, AND ASKED THAT THE ENVIRONMENTAL PROTECTION AGENCY BE REQUIRED TO PROMULGATE IMMEDIATE REGULATIONS CONTROLLING SUCH COMPLEX SOURCES OF AIR POLLUTION.

THE SUIT WAS DISMISSED BY U.S. DISTRICT COURT FOR THE CENTRAL DISTRICT OF CALIFORNIA, WHICH HELD THAT SECTION 304 OF THE ACT AUTHORIZES CITIZEN SUITS ONLY IN CASES OF VIOLATION OF EXISTING STANDARDS. NO REGULATION PRESENTLY EXISTS TO CONTROL SHOPPING CENTER EMISSIONS.

THE APPEALS COURT, IN AN OPINION BY JUDGE ROBERT H. SCHNAKE SAID, "THE AREA OF THE SHOPPING CENTER HERE INVOLVED IS PRESENTLY SUBJECT TO NO CONTROL THAT WOULD PREVENT ITS CONSTRUCTION OR OPERATION."

SCHNAKE SAID THAT CERTAIN TYPES OF ACTION BY EPA MAY BE ORDERED BY THE COURT, BUT EMPHASIZED THAT THE COURT "CANNOT ORDER THE PROMULGATION OF ANY PARTICULAR REGULATION OR SET OF REGULATIONS." ONCE SUCH REGULATIONS ARE PROMULGATED, REVIEW LIES IN THE FEDERAL COURTS OF APPEALS PURSUANT TO SECTION 307 RATHER THAN A FEDERAL DISTRICT COURT UNDER SECTION 304.

REFERENCE	ENVIRONMENT REPORTER, PUBLISHED BY THE BUREAU OF NATIONAL AFFAIRS, INC., VOL. 5 NO. 13, JULY 26, 1974, PP. 400-401

FOR THE FULL TEXT OF THE OPINION, SEE:
ENVIRONMENT REPORTER CASES, VOL. 6, P. 1606.

CASE	PLAN JEFF CO. VS. KLEPPE, NO. 76-405
LOCATION	DENVER, COLORADO
DATE & STATUS	DISMISSED JUNE 25, 1976
PROJECT TYPE	DEVELOPMENT OF OFFICE SPACE
ISSUE(S)	NEPA APPLICABILITY, ISSUANCE OF LEASES, TRAFFIC, AIR AND NOISE POLLUTION, ECONOMIC DECAY, SOCIAL IMPACT
SCOPE	IN AN ACTION TO RESTRAIN THE GENERAL SERVICE ADMINISTRATION FROM GRANTING LEASES PERMITTING PRIVATE DEVELOPMENT OF OFFICE SPACE TO BE OCCUPIED BY FEDERAL AGENCY PERSONNEL IN THE AREA SURROUNDING THE DENVER FEDERAL CENTER, THE OPPOSITION CONTENDED THAT PREPARATION OF AN EIS IS REQUIRED BEFORE BUILDINGS CAN BE OCCUPIED, SINCE APPROVAL OF LEASES CONSTITUTES A MAJOR FEDERAL ACTION. PLAINTIFFS FURTHER CONTENDED THAT UNRESTRICTED DEVELOPMENT OF THE AREA WILL RESULT IN TRAFFIC CONGESTION, AIR AND NOISE POLLUTION, AND SUBSTANTIAL NEIGHBORHOOD DECAY. ON JUNE 25, 1976, THE PARTIES STIPULATED TO DISMISS THE COMPLAINT WITHOUT PREJUDICE. THE PARTIES AGREED THAT FUTURE LEASES OF THIS TYPE WILL NOT BE AWARDED UNTIL A FINAL EIS IS ISSUED FOR THE FEDERAL CENTER REGION AND AN ANALYSIS OF THE CUMULATIVE IMPACT OF FEDERAL LEASING IN THE AREA IS PREPARED.
REFERENCE	ENVIRONMENTAL LAW REPORTER, DIGEST, VOL. 2, P 65363

CASE	PRINCE GEORGE'S COUNTY VS. HOLLOWAY, NO. 75-1823
LOCATION	WASHINGTON, D.C.
DATE & STATUS	ENJOINED ON DECEMBER 5, 1975
PROJECT TYPE	RELOCATING OCEANOGRAPHIC CENTER
ISSUE(S)	EIS ADEQUACY, HIGH DENSITY DEVELOPMENT

SCOPE

THE COURT ENJOINED THE U.S. NAVY FROM TRANSFERRING ITS OCEANOGRAPHIC CENTER FROM A SUBURB OF WASHINGTON, D.C. TO BAY ST. LOUIS, MISSISSIPPI, UNTIL A NEW EIS HAD BEEN PREPARED.

THE PLAINTIFF CHALLENGED THE SUFFICIENCY OF THE NAVY'S COMPLIANCE WITH NEPA, STATING THAT THE PRESENT EIS FAILS TO COMPLY WITH NEPA IN THAT IT DOES NOT CONSIDER THE POSSIBLE CUMULATIVE IMPACT OF EXISTING ARMY AND NASA PROPOSALS FOR BUILDING AN AMMUNITION PLANT AND EXPANDING THE SPACE SHUTTLE PROGRAM AT THE SAME BAY ST. LOUIS SITE. ALSO, THEY ARGUED THAT THE EIS FAILED TO EXAMINE POSSIBLE METHODS OF MITIGATING THE PLAIN PROBLEM WHICH THE MOVE WILL ENTAIL RESPECTING AVAILABILITY OF ADEQUATE HOUSING AND SCHOOLS FOR LOW AND MODERATE INCOME GROUPS AND RACIAL MINORITIES AMONG THE CENTER'S EMPLOYEES.

THE DEFENDANTS ARGUED THAT AFTER CONSIDERING 15 OTHER POSSIBLE SITES, THE SITE AT BAY ST. LOUIS, MISSISSIPPI, WHERE NASA COULD OFFER UNUTILIZED FACILITIES RELEASED WITH THE TERMINATION OF THE APOLLO MOON PROGRAM, WAS THE MOST DESIRABLE FOR REASONS OF EFFICIENCY AND COST. THE DEFENDANTS ADMITTED THAT THEY WERE WELL AWARE OF THE NEPA AND, AFTER PUBLIC HEARINGS, ISSUED A FINAL EIS WHICH INVESTIGATED THOROUGHLY THE ENVIRONMENTAL EFFECT OF THE MOVE AT BAY ST. LOUIS.

THE COURT CONCLUDED THAT THE PROPOSED ACTIONS BY THE ARMY AND BY NASA SHOULD HAVE BEEN CONSIDERED IN THE NAVY'S EIS, AS THE PROPOSALS HAD BEEN ADVANCED BEYOND THE POINT OF CONJECTURE AND SPECULATION. THUS, IT WAS APPARENT TO THE COURT ON THE PRESENT RECORD THAT THERE WAS A STRONG POSSIBILITY THAT THESE GOVERNMENTAL UNITS WOULD ALSO BE OCCUPYING THE SITE AT BAY ST. LOUIS TOGETHER WITH THE NAVY. THE NEPA, THEN, REQUIRED THE IMPACT STATEMENT TO CONSIDER THE CUMULATIVE ENVIRONMENTAL EFFECT THAT THE ACTIONS MAY HAVE AT THE COMMON SITE. ONE OF THE PRIMARY PURPOSES OF THE ACT WAS TO PREVENT THE VERY TYPE OF FRAGMENTED ANALYSIS THAT OCCURRED HERE. INSTEAD, THE STATUTE DIRECTED THAT THE AGENCY EMPLOY A MORE INTEGRATED AND COMPREHENSIVE APPROACH WHICH TAKES ACCOUNT OF THE OVERALL EFFECT OF THE VARIOUS PROJECTS. ALSO, THE RELOCATION OF THE OCEANOGRAPHIC SITE RAISED QUESTIONS OF ADEQUATE HOUSING AND SCHOOLING WHICH ARE OF MAJOR ENVIRONMENTAL IMPORTANCE. THUS, THE COURT RULED THAT NO RELOCATION SHALL ENSUE UNTIL THE SECRETARY OF DEFENSE PREPARES AN ADEQUATE EIS.

REFERENCE: ENVIRONMENTAL LAW REPORTER, VOL. 6, 1975, PP. 20109-20111

CASE	SAN FRANCISCO TOMORROW VS. ROMNEY, C-72-65 RHS
LOCATION	SAN FRANCISCO AND BERKELEY, CALIFORNIA
DATE & STATUS	REINSTATED ON APRIL 24, 1972
PROJECT TYPE	URBAN RENEWAL PROJECTS
ISSUE(S)	NEPA APPLICABILITY, FEASIBILITY OF A PROJECT
SCOPE	THE U.S. DISTRICT COURT RULED THAT THE PLAINTIFFS DO NOT HAVE STANDING UNDER NEPA TO CHALLENGE HUD'S FINANCIAL ASSISTANCE TO TWO URBAN RENEWAL PROJECTS, THE YERBA BUENA PROJECT IN SAN FRANCISCO AND A SIMILAR, LESS EXTENSIVE PROJECT IN BERKELEY. SUIT WAS BROUGHT BY SAN FRANCISCO TOMORROW, A GROUP OF ORGANIZATIONS ALLEGING GENERAL CONCERN FOR THE CONDITION OF THE ENVIRONMENT AND THE WELFARE OF SOCIETY, AND SEVERAL INDIVIDUALS WHO CLAIM RESIDENCE IN THE VICINITY OF THE PROJECTS INVOLVED. THEY CHARGED THAT THE TWO PROJECTS ARE UNFEASIBLE, SUPPLYING THE COURT WITH AFFIDAVITS FROM GEOLOGISTS, ENGINEERS, AND CITY PLANNERS. IN ITS OPINION, THE COURT NOTED THAT NOTHING IN NEPA SUGGESTS THAT FEASIBILITY OR THE LACK THEREOF IS THE PROPER CONCERN OF THE COURT. THE PLAINTIFFS, THE COURT RULED, CLAIM NO PROPERTY OR OTHER LEGAL INTEREST IN THE PROJECTS DIFFERENT FROM THAT POSSESSED BY THE CITIZENRY AT LARGE. PERSONS WHO HAVE A "MERE NON-PECUNIARY INTEREST" IN THE ENFORCEMENT OF NEPA HAVE NO STANDING TO SUE UNDER THAT STATUTE. MOREOVER, THE COURT SAID, HUD HAD NO OBLIGATION TO COMPLY WITH NEPA BECAUSE ALL RELEVANT PLANNING OF THE PROJECTS HAD BEEN DETERMINED PRIOR TO THE EFFECTIVE DATE OF THE STATUTE, AND ANY CHANGES IN THE PLANS SINCE THAT DATE WILL NOT HAVE A SIGNIFICANT EFFECT ON THE ENVIRONMENT.
REFERENCE	ENVIRONMENTAL LAW REPORTER, VOL. 2, PP. 20273-20275

CASE	SAVE OUR TEN ACRES VS. GENERAL SERVICES ADMINISTRATION
LOCATION	MOBILE, ALA.
DATE & STATUS	
PROJECT TYPE	FEDERAL OFFICE BUILDING
ISSUE(S)	NEPA REQUIREMENTS, EFFECT ON PARKING, TRAFFIC CONGESTION AND AIR POLLUTION
SCOPE	A COURT CAN REVIEW A FEDERAL AGENCY'S DECISION THAT A PROPOSED PROJECT DOES NOT SIGNIFICANTLY AFFECT THE ENVIRONMENT WITHOUT ALLEGATIONS THAT THE PROJECT WOULD MATERIALLY DEGRADE ANY ASPECT OF THE ENVIRONMENTAL QUALITY, THE U.S. COURT OF APPEALS FOR THE FIFTH CIRCUIT RULED. THE SUIT TO ENJOIN CONSTRUCTION WAS BROUGHT BY SAVE OUR TEN ACRES WHO CHARGED THAT THE FAILURE OF THE GENERAL SERVICES ADMINISTRATION TO PREPARE AN ENVIRONMENTAL IMPACT STATEMENT VIOLATED NEPA. GSA CLAIMED THAT THE PROJECT WAS NOT A MAJOR FEDERAL ACTION SIGNIFICANTLY AFFECTING THE ENVIRONMENT. THE DISTRICT COURT HELD THAT THE INITIAL GSA DETERMINATION AS TO THE BUILDING'S ENVIRONMENTAL EFFECT COULD NOT BE DISTURBED UNLESS IT WAS FOUND TO BE ARBITRARY, CAPRICIOUS OR AN ABUSE OF DISCRETION. THE APPELLATE COURT REVERSED, HOLDING THAT AN AGENCY'S DECISION AS TO WHETHER A PROJECT IS A MAJOR FEDERAL ACTION MUST BE SUBJECT TO "A MORE SEARCHING STANDARD" OF REVIEW THAN THE ARBITRARY AND CAPRICIOUS STANDARD. THE COURT DETERMINED THAT THE ALLEGATIONS OF SAVE OUR TEN ACRES CONCERNING THE BUILDING'S EFFECT ON PARKING, TRAFFIC CONGESTION, AND AIR POLLUTION WERE SUFFICIENT TO ALLOW A COURT TO WEIGH THE EVIDENCE TO DETERMINE WHETHER THE AGENCY'S CONCLUSION WAS RESSONABLE. THE COURT FURTHER NOTED THAT THIS JUDICIAL REVIEW NEED NOT BE LIMITED TO A CONSIDERATION OF THE ADMINISTRATIVE RECORD BUT CAN, IF AN "INADEQUATE EVIDENTIARY DEVELOPMEMT" BEFORE THE AGENCY IS SHOWN, ENCOMPASS SUPPLEMENTARY AFFIDAVITS AND PROOFS CONCERNING THE BUILDING'S ENVIRONMENTAL IMPACT. THE APPEALS COURT REMANDED THE CASE TO THE U.S. DISTRICT COURT FOR THE SOUTHERN DISTRICT OF ALABAMA.
REFERENCE	ENVIRONMENT REPORTER, PUBLISHED BY THE BUREAU OF NATIONAL AFFAIRS, INC., VOL. 3, NO. 39, JAN. 26, 1973, P. 1165

CASE	SAVE THE COURTHOUSE COMMITTEE VS. LYNN, NO. 74-CIV.5646
LOCATION	WHITE PLAINS, NEW YORK
DATE & STATUS	ENJOINED MARCH 6, 1976
PROJECT TYPE	URBAN RENEWAL PROJECT
ISSUE(S)	NEPA, NHPA APPLICABILITY
SCOPE	THE PLAINTIFFS ALLEGED SUFFICIENT HARM TO THEIR INTERESTS OF PRESERVING A COURTHOUSE, PART OF AN URBAN RENEWAL PLAN UNDER HUD. THEY PROTESTED THAT THEIR INTEREST IS PROTECTED UNDER THE NEPA AND THE NATIONAL HISTORIC PRESERVATION ACT. THE COURT FOUND THAT HUD FAILED ON THREE COUNTS: 1. TO OBSERVE THE REGULATION REQUIRING AGENCY DEPARTMENT HEADS TO IDENTIFY PROPERTIES WITHIN THE AREA OF THEIR UNDERTAKINGS THAT ARE ELIGIBLE FOR INCLUSION ON THE NATIONAL REGISTER; 2. TO DETERMINE PROJECT EFFECTS ON SUCH PROPERTIES, AND IF ADVERSE IMPACT IS FOUND; 3. TO REQUEST COMMENTS FROM THE ADVISORY COUNCIL IN WRITING. THE COURT HELD THAT PLAINTIFFS ESTABLISHED IRREPARABLE HARM FROM THE PROPOSED DEMOLITION. THUS, IN CONJUNCTION WITH COMPLIANCE WITH NEPA AND NHPA, PLAINTIFFS' HARM OUTWEIGHED THE FINANCIAL HARM THAT AN INJUNCTION WOULD IMPOSE ON THE DEFENDANTS. THE COURT EXTENDED THE INJUNCTION TO THE LOCAL WHITE PLAINS REDEVELOPMENT AGENCY WHICH IS IN A CLOSE AND CONTINUING PARTNERSHIP WITH HUD IN IMPLEMENTING THIS URBAN RENEWAL PROJECT.
REFERENCE	ENVIRONMENTAL LAW REPORTER, VOL. 6, P 20639

CASE	SCENIC RIVERS ASSOCIATION VS. LYNN, NO. 74-131-C
LOCATION	ILLINOIS
DATE & STATUS	ENJOINED SEPTEMBER 4, 1974
PROJECT TYPE	HOUSING DEVELOPMENT
ISSUE(S)	NEPA REQUIREMENTS, WATER QUALITY
SCOPE	PLAINTIFFS CONTENDED THAT THERE IS THREAT OF INJURY TO THE PUBLIC INTEREST IN THE PRESERVATION OF THE ILLINOIS RIVER BASIN IF THE OFFICE OF INTERSTATE LAND SALES OF HUD DOES NOT COMPLY WITH NEPA'S PROCEDURES. THE COURT ENJOINED THE OFFICE FROM APPROVING THE DEVELOPER'S REPORT FOR A PLANNED DEVELOPMENT OF 3,000 HOMES ALONG THE ILLINOIS RIVER UNTIL A NEPA EIS HAS BEEN PREPARED. THE COURT FOUND THAT THE PLAINTIFFS HAD ADEQUATELY SHOWN THAT THE DEVELOPMENT'S SEPTIC TANKS HAVE THE POTENTIAL FOR AFFECTING THE QUALITY OF THE HUMAN ENVIRONMENT BY POLLUTING THE ILLINOIS RIVER. THE OFFICE, THEREFORE, IS ORDERED TO WITHDRAW ITS PREVIOUS APPROVAL OF DEVELOPMENT, AND PROHIBITED FROM REINSTATING THAT APPROVAL PENDING A FURTHER ORDER OF THE COURT.
REFERENCE	ENVIRONMENTAL LAW REPORTER, VOL. 4, PP 20727-8

CASE	SIBSON VS. STATE, EQUITY NO. 4815
LOCATION	NEW HAMPSHIRE
DATE & STATUS	ENJOINED SEPTEMBER 24, 1973
PROJECT TYPE	HOUSING DEVELOPMENT
ISSUE(S)	LAND FILLING PERMIT, PRIVATE PROPERTY RIGHTS
SCOPE	THE PLAINTIFF LAND OWNERS REQUESTED PERMISSION TO FILL FOUR ACRES OF TIDAL MARSH ON THEIR PROPERTY FOR THE PURPOSE OF DEVELOPING HOUSES. IN APPLYING FOR THE PERMIT THE WATER RESOURCES BOARD DENIED THE REQUEST. THE PLAINTIFFS ARGUED THAT THIS CONSTITUTED UNJUST TAKING OF PLAINTIFFS' PROPERTY. THE COURT UPHELD THE BOARD'S DECISION STATING THAT THE PLAINTIFFS FAILED TO SUPPORT THEIR BURDEN OF PROVING THE BOARD'S ACTION TO BE UNREASONABLE OR UNLAWFUL. THE COURT FOUND THAT THE DENIAL WAS A VALID EXERCISE OF THE POLICE POWER, RATHER THAN A CONFISCATION OF PRIVATE PROPERTY, SINCE THE MARSH IS A VALUABLE PART OF THE STATE'S ENVIRONMENT AND WOULD BE DESTROYED BY THE FILLING OPERATION. THE COURT ALSO HELD THAT THOUGH THE ACTION DEFLATED THE PROPERTIES' SPECULATIVE VALUE, PROFIT ON THE INITIAL PURCHASE HAD ALREADY BEEN MADE BY BUILDING AND SELLING A HOUSE ON A NON-TIDAL PORTION OF THE TRACT. THUS, THE COURT DISMISSED THE PLAINTIFFS' PETITIONS AND ENJOINED THE HOUSING DEVELOPMENT.
REFERENCE	ENVIRONMENTAL LAW REPORTER, VOL. 4, P 20060

CASE	SIERRA CLUB VS. LYNN, NO. 73-3378
LOCATION	SAN ANTONIO, TEXAS
DATE & STATUS	REINSTATED OCTOBER 4, 1974
PROJECT TYPE	RANCH PROJECT (RESIDENTIAL DEVELOPMENT)
ISSUE(S)	EIS ADEQUACY, WATER POLLUTION
SCOPE	AN OFFER OF COMMITMENT BY HUD TO GUARANTEE AN 18 MILLION DOLLAR PRIVATE BOND ISSUE FOR THE DEVELOPMENT OF SAN ANTONIO RANCH NEW TOWN (RANCH) SPAWNED ENVIRONMENTALIST LITIGATION AGAINST DEFENDANTS, HUD SECRETARY AND THE DEVELOPER. PLAINTIFFS SOUGHT AN APPROPRIATE EIS FROM THE DEFENDANTS BECAUSE THEY FELT THAT THE DEVELOPMENT THREATENS TO POLLUTE UNDERGROUND WATER SUPPLY. THE COURT OF APPEALS AFFIRMED THE LOWER COURT'S DECISION THAT HUD'S EIS FOR THE DEVELOPMENT CONFORMED IN ALL WAYS TO THE REQUIREMENTS OF NEPA, THOUGH THE ORIGINAL EIS DID NOT.
REFERENCE	ENVIRONMENTAL LAW REPORTER, VOL. 4, PP 20844-20854

CASE	STATE VS. ASHMORE, NOS. 30176, 30177
LOCATION	GEORGIA
DATE & STATUS	ENJOINED FEBRUARY 24, 1976
PROJECT TYPE	CONDOMINIUM CONSTRUCTION
ISSUE(S)	BOUNDARIES OF LANDS ADJACENT TO NAVIGABLE TIDEWATERS
SCOPE	THE COURT UPHELD THE CONSTITUTIONALITY OF A 1902 GEORGIA STATUTE THAT EXTENDED TO THE LOW WATER MARK THE BOUNDARIES OF LANDS ADJACENT TO NAVIGABLE TIDEWATERS. THE 1902 ACT WAS IMMUNIZED FROM FURTHER CONSTITUTIONAL ATTACK BY EXPRESS RATIFICATION IN THE 1975 STATE CONSTITUTION. THE ACT DOES NOT GRANT FEE SIMPLE TITLE TO THIS FORESHORE TO THE RIPARIAN OWNERS. A PORTION OF THE ACT GRANTING TITLE TO OWNERS ADJACENT TO NON-NAVIGABLE TIDEWATERS DOES NOT APPLY TO THIS CONTROVERSY. THE DISSENT ARGUED THAT THE 1902 ACT WAS DESIGNED TO REMEDY THE RESTRICTIVE EFFECT OF THE COURT'S PRIOR DECISION, AND THIS SHOULD BE READ TO GRANT TITLE TO THE ADJACENT OWNERS BY EXTENDING THE BOUNDARIES OF THE FORESHORE. FURTHERMORE, THE MAJORITY OPINION ALLOWS PRIVATE PROPERTY TO BE TAKEN WITHOUT JUST COMPENSATION. A SECOND DISSENT INTERPRETS A STATUTORY CAVEAT TO THE ACT'S BOUNDARY EXTENSION AS CREATING AN EASEMENT FOR PUBLIC USE OF THE FORESHORE FOR FISHING, PASSAGE, AND TRANSPORTATION, NOT, AS APPELLEES CONTENDED, AS LIMITED TO COMMON LAW PUBLIC TRUST USES OF NAVIGATION AND COMMERCE OVER THE WATER COVERING THE FORESHORE. THE DEFENDANT PLANNED TO CONSTRUCT CONDOMINIUMS ON THE ACCRETED LANDS WHICH ACCRUE TO THE SUBDIVISION LOT OWNERS BORDERING THE SUBDIVISION BOUNDARY. THE COURT CONCLUDED THAT THE APPELLEES' LAND, BOUNDED BY THE ATLANTIC OCEAN, HAS AS SEAWARD BOUNDARY THE LOW-WATER MARK. HOWEVER, THE APPELLEES' LAND THAT LIES WITHIN THE FORESHORE IS SUBJECT TO AN EASEMENT IN FAVOR OF THE PUBLIC AS DELINEATED IN THIS DISSENTING OPINION. ALSO, THE PEOPLE HAVE NO RIGHTS IN ACCRETED LAND OUTSIDE THE CONFINES OF THE SHIFTING FORESHORE, UNLESS SUCH PUBLIC RIGHTS HAVE BEEN ACQUIRED BY CONVEYANCE OR PRESCRIPTION USE. THUS, THE STATE HOLDS TITLE TO THE FORESHORE ALONG NAVIGABLE TIDEWATERS.
REFERENCE	ENVIRONMENTAL LAW REPORTER, VOL. 6, 1976, PP. 20438-20444

CASE	STATE HIGHWAY COMMISSION VS. BAUMAN, NO. 28831
LOCATION	OREGON
DATE & STATUS	REINSTATED JANUARY 21, 1974
PROJECT TYPE	CONDOMINIUM PROJECT
ISSUE(S)	PUBLIC RECREATIONAL RIGHTS, PRIVATE PROPERTY RIGHTS
SCOPE	PLAINTIFF SOUGHT REVERSAL OF A LOWER COURT RULING DENYING DECLARATORY RELIEF AGAINST DEFENDANTS WHO PLAN TO CONSTRUCT A CONDOMINIUM PROJECT ON A COASTAL SAND DUNE IN WHICH, PLAINTIFF CLAIMED, THE PUBLIC HAS ACQUIRED RECREATION RIGHTS BY PRESCRIPTION, DEDICATION, OR CUSTOM. THE COURT AFFIRMED THE LOWER COURT RULING DENYING DECLARATORY OR INJUNCTION RELIEF ON THE GROUNDS THAT THE PLAINTIFF FAILED TO PROVE THAT THE PUBLIC HAD ACQUIRED RECREATIONAL RIGHTS BY DEDICATION. AN INTENT TO DEDICATE LAND TO PUBLIC USE MUST BE CLEARLY AND UNEQUIVOCALLY MANIFESTED IN THE OWNER'S ACTIONS IN ORDER FOR A COURT TO FIND THAT AN IMPLIED DEDICATION HAS OCCURRED. IN THIS CASE, THE COURT FOUND NO EVIDENCE OF SUCH AN INTENT. THE COURT ALSO POINTED OUT THAT THE PLANS FOR THE PROJECT INCLUDE PROVISIONS FOR SEVERAL ACCESS ROUTES OVER THE PROPERTY TO THE PUBLIC BEACH, ALLOWING FREE PASSAGE BY THE PUBLIC.
REFERENCE	ENVIRONMENTAL LAW REPORTER, VOL. 4, P 20311

CASE	STEWART VS. DISTRICT OF COLUMBIA BOARD OF ZONING ADJUSTMENT, NO. 6802
LOCATION	WASHINGTON, D.C.
DATE & STATUS	ENJOINED ON MAY 16, 1973
PROJECT TYPE	ATHLETIC CLUB AND RECREATIONAL FACILITY
ISSUE(S)	COMMUNITY FACILITY ZONING REQUIREMENTS
SCOPE	THE COURT REVERSED THE DECISION OF THE DISTRICT OF COLUMBIA BOARD OF ZONING ADJUSTMENT WHICH GRANTED A SPECIAL EXCEPTION ALLOWING THE CONSTRUCTION OF A PRIVATE TENNIS CLUB ON OPEN SPACE IN A DISTRICT ZONED FOR SINGLE-FAMILY DETACHED DWELLINGS. WHERE A TENNIS AND RECREATIONAL FACILITY IS NOT OPEN TO THE COMMUNITY AT LARGE DUE TO MEMBERSHIP REQUIREMENTS AND LIMITATIONS, AND IS NOT ESTABLISHED AS A PURELY LOCAL FACILITY, THE COURT SAID, THERE IS NO BASIS FOR A DECISION BY THE ZONING BOARD THAT IT IS A COMMUNITY FACILITY ENTITLED TO A SPECIAL EXCEPTION UNDER THE ORDINANCE. THE PROPOSED BUILDINGS WERE TO HOUSE AREAS FOR INDOOR TENNIS, SQUASH, HANDBALL, SAUNA BATHS, AND INDOOR SWIMMING. THE ORDER BY THE BOARD OF ZONING ADJUSTMENT PERMITTED THE OPERATION OF THE FACILITIES AS A CLUB FOR ITS MEMBERS AND THEIR GUESTS ONLY. THE ZONING REGULATION'S REQUIREMENT, THE COURT RULED, IS THAT THE PROPOSED FACILITY BE A COMMUNITY CENTER FACILITY OPERATED BY A LOCAL COMMUNITY ORGANIZATION. THE BOARD WAS WITHOUT POWER TO GRANT AN EXCEPTION TO A PRIVATE CLUB, AND THE COURT REVERSED ITS ORDER.
REFERENCE	ENVIRONMENTAL LAW REPORTER, VOL. 3, PP 20477-20479

CASE	SWEETWATER VALLEY CIVIC ASSOCIATION VS. CITY OF NATIONAL CITY, NO. 14408
LOCATION	CALIFORNIA
DATE & STATUS	FILED AUGUST, 1975
PROJECT TYPE	CONSTRUCTION OF REGIONAL SHOPPING CENTER
ISSUE(S)	INADEQUATE EIS, ALTERNATIVE ACTIONS
SCOPE	STATE OF CALIFORNIA SOUGHT REVERSAL OF A LOWER COURT DECISION DENYING PETITION FOR MANDATORY INJUNCTION TO PREVENT A MUNICIPAL REDEVELOPMENT AGENCY'S DESIGNATION OF THE BONITA GOLF COURSE AS "BLIGHTED" THEREBY SUBJECTING IT TO REDEVELOPMENT AS THE SITE FOR CONSTRUCTION OF A REGIONAL SHOPPING CENTER. THE PLAINTIFF ARGUED THAT SUCH DESIGNATION IS AN ABUSE OF DISCRETION IN THAT IT REQUIRES A DETERMINATION OF ECONOMIC, ENVIRONMENTAL, OR SOCIAL DECLINE, NOT PRESENT IN THE CASE OF THE BONITA GOLF COURSE. ALSO THEY ARGUED THAT SUCH DESIGNATION IS AIMED PURELY AT INCREASING THE CITY'S NET TAX REVENUES AND CONFLICTS WITH THE STATED LEGISLATIVE POLICY OF BOTH CALIFORNIA AND THE AGENCY TO PRESERVE AND PROTECT PRODUCTIVE OPEN SPACES AND ENVIRONMENTALLY SENSITIVE AREAS. THE PLAINTIFF ALSO HELD THAT THE ENVIRONMENTAL IMPACT REPORT FOR THE REDEVELOPMENT IS LEGALLY INSUFFICIENT BECAUSE IT DOES NOT PROVIDE THE REQUIRED DETAILED STATEMENT OF ALTERNATIVES TO THE PROPOSED DEVELOPMENT PLAN.
REFERENCE	ENVIRONMENTAL LAW REPORTER, VOL. 6, P 65261

CASE	THREE LAKES ASSOCIATION VS. FISHER, NO. 1142
LOCATION	TORCH LAKE, MICHIGAN
DATE & STATUS	REINSTATED AUGUST 20, 1973
PROJECT TYPE	HOUSING PROJECTS
ISSUE(S)	POLLUTION AND OVERCROWDING OF THE LAKE
SCOPE	PROPERTY OWNERS ON AND NEAR TORCH LAKE, IN ANTRIM COUNTY, CLAIMED THAT PROPOSED HOUSING PROJECTS WOULD RESULT IN BOTH POLLUTION AND OVERCROWDING OF THE LAKE. DEFENDANTS PLANNED TO BUILD HOUSES ON A SMALL RIPARIAN TRACT AT THE LAKE AND A MUCH LARGER ADJOINING BACK TRACT, GIVING LAKE ACCESS TO ALL THE OWNERS. THE CLAIM BY THE PLAINTIFFS QUESTIONED THE EXTENT TO WHICH RIPARIAN OWNERS' TRADITIONAL OPPORTUNITIES FOR WATER ACCESS COULD BE "STRETCHED" TO INCLUDE OWNERS OF BACK LOTS. THE COURT NEGOTIATED A CONSENT JUDGEMENT PROVIDING A COMPROMISE WHICH REDUCED THE NUMBER OF BACK LOTS, PROVIDED A POOL AND OTHER RECREATIONAL FACILITIES ON THE ACCESS TRACT TO REDUCE RECREATIONAL DEMAND ON THE LAKE, RESTRICTED THE USE OF FERTILIZERS, AND RESTRICTED THE USE OF THE LAKE BY BACK LOT OWNERS.
REFERENCE	ECOLOGY LAW QUARTERLY, VOL. 4:1, 1974, P. 10

CASE	TIERRASANTA COMMUNITY COUNCIL VS. RICHARDSON, NO. 73-277-E
LOCATION	SAN DIEGO, CALIFORNIA
DATE & STATUS	ENJOINED NOVEMBER 6, 1974
PROJECT TYPE	JUVENILE DETENTION FACILITY
ISSUE(S)	NEPA REQUIREMENTS, ZONING CODES
SCOPE	THE PLAINTIFFS MOVED FOR A PRELIMINARY INJUNCTION ON THE CONSTRUCTION OF A $4.5 MILLION JUVENILE DETENTION FACILITY NEAR A DESIGNATED ELEMENTARY SCHOOL SITE IN A PLANNED, LOW DENSITY RESIDENTIAL AREA, ON THE GROUNDS THAT THE PROPOSED FACILITY CONSTITUTES A MAJOR FEDERAL ACTION HAVING SIGNIFICANT PSYCHOLOGICAL AND SOCIOLOGICAL EFFECTS ON THE COMMUNITY. THEY REQUESTED THAT A NEPA IMPACT STATEMENT BE FILED. THE COURT AGREED THAT WHERE THE CONSTRUCTION WILL ARGUABLY HAVE A SIGNIFICANT EFFECT ON THE HUMAN ENVIRONMENT, IT IS WISER TO ORDER PREPARATION OF AN EIS AND THUS AVOID THE DELAY AND UNCERTAINTY OF TRIAL. A NEGATIVE IMPACT DETERMINATION CONCERNING A PROPOSED FACILITY WHICH DEVIATES FROM LOCAL ZONING PROCEDURES MUST BE SUBJECTED TO PARTICULAR SCRUTINY; AN EIS WOULD ENABLE SUCH SCRUTINY. THUS, THE COURT PRELIMINARILY ENJOINED THE GENERAL SERVICE ADMINISTRATION FROM BEGINNING CONSTRUCTION OF THE JUVENILE DETENTION FACILITY.
REFERENCE	ENVIRONMENTAL LAW REPORTER, VOL. 4, P 20309

CASE	TRINITY EPISCOPAL SCHOOL VS. ROMNEY, NOS. 75-7061, 7092
LOCATION	NEW YORK, NEW YORK
DATE & STATUS	ENJOINED ON 24 JULY, 1975
PROJECT TYPE	PUBLIC HOUSING PROJECT
ISSUE(S)	NEPA REQUIREMENTS, ALTERNATIVE ACTIONS, ALTERNATIVE SITES

SCOPE

THE U.S. COURT OF APPEALS FOR THE SECOND CIRCUIT RULED THAT SECTION 102 (2) (D) OF THE NEPA REQUIRES FEDERAL AGENCIES TO CONSIDER ALTERNATIVES TO PROPOSED AGENCY ACTION REGARDLESS OF WHETHER THAT ACTION REQUIRES PREPARATION OF AN ENVIRONMENTAL IMPACT STATEMENT UNDER SECTION 102 (2) (C).

TRINITY EPISCOPAL SCHOOL, A PRIVATE SCHOOL LOCATED ON THE WEST SIDE OF NEW YORK CITY, BROUGHT SUIT AGAINST HUD AND STATE AND LOCAL HOUSING AGENCIES. TRINITY CLAIMED THAT HUD VIOLATED NEPA IN AWARDING FEDERAL FUNDS FOR DEVELOPMENT OF A PUBLIC HOUSING PROJECT, WHICH IS PART OF THE OVERALL WEST SIDE URBAN RENEWAL PLAN.

HUD PREPARED A "SPECIAL ENVIRONMENTAL CLEARANCE" CONCERNING THE PUBLIC HOUSING PROJECT AND CONCLUDED THAT THE PROJECT WOULD NOT HAVE A SIGNIFICANT ADVERSE AFFECT UPON THE ENVIRONMENT AND THAT THEREFORE PREPARATION OF AN IMPACT STATEMENT WAS NOT REQUIRED.

IN AN OPINION BY JUDGE LEONARD P. MOORE, THE COURT RULED THAT "FEDERAL AGENCIES MUST CONSIDER ALTERNATIVES UNDER SECTION 102 (2) (D) WITHOUT REGARD" TO THE FILING OF AN IMPACT STATEMENT, AND THAT HUD WAS THEREFORE IN VIOLATION OF THIS SECTION.

THE COURT FOUND THAT HUD SIMPLY ADOPTED THE CONCLUSION OF THE NEW YORK CITY HOUSING AUTHORITY THAT NO ALTERNATIVE SITES ARE AVAILABLE DUE TO THE SCARCITY OF LAND. "THIS BLANKET GENERALIZATION UNSUPPORTED BY THE EVIDENCE AS TO ALTERNATIVE SITES WITHIN THE AREA (NOT TO MENTION OTHER POTENTIAL SITES) DOES NOT CONFORM WITH HUD'S RESPONSIBILITIES," THE COURT SAID.

"HUD IS NOT REQUIRED TO SEARCH OUT POTENTIAL SITES THROUGHOUT THE NEW YORK METROPOLITAN AREA BUT THE FEDERAL AGENCY MUST ITSELF DETERMINE WHAT IS REASONABLY AVAILABLE ESPECIALLY WHERE AS HERE THE BUILDING OF PUBLIC HOUSING IS PART OF A COORDINATED PLAN TO DEAL WITH THE BROAD PROBLEM OF MEETING THE HOUSING NEEDS OF LOW INCOME RESIDENTS," ACCORDING TO THE COURT.

ALTERNATIVES THAT MUST BE CONSIDERED BY HUD, THE COURT SAID, INCLUDE THE ALTERNATIVE OF NOT BUILDING THE PROJECT, USE OF ALTERNATIVE DESIGNS, DISPERSAL OF THE PUBLIC HOUSING UNITS ON MORE SITES WITHIN THE WEST SIDE PROJECT AREA, AND ALTERNATIVES "REQUIRING ACTION OF A SIGNIFICANTLY DIFFERENT NATURE" SUCH AS REHABILITATION OF EXISTING BUILDINGS FOR PUBLIC HOUSING.

REFERENCE

ENVIRONMENT REPORTER, PUBLISHED BY THE BUREAU OF NATIONAL AFFAIRS, INC., VOL. 6, NO. 15, AUG. 8, 1975, P. 585

CASE	ULSTER COUNTY COMMUNITY ACTION COMMITTEE VS. KOENIG, NO. 75-CIV.-3832
LOCATION	KINGSTON, N.Y.
DATE & STATUS	REINSTATED ON OCTOBER 17, 1975
PROJECT TYPE	COMMUNITY DEVELOPMENT PLAN
ISSUE(S)	IMPACT STATEMENT REQUIREMENT AS APPLIED TO GRANT APPLICATION
SCOPE	A FEDERAL DISTRICT COURT REFUSED TO ENJOIN THE KINGSTON, N.Y. COMMUNITY DEVELOPMENT PLAN, PROPOSED IN APPLICATION FOR FEDERAL ASSISTANCE UNDER THE HOUSING AND COMMUNITY DEVELOPMENT ACT, ON GROUNDS THAT CITY OFFICIALS DID NOT FILE A NEPA ENVIRONMENTAL IMPACT STATEMENT, ALSO RULING THAT THE PLAN CONFORMS TO GRANT APPLICATION REQUIREMENTS OF THE HOUSING ACT. THE CITIZENS GROUP BRINGING THE SUIT CONTENDED THAT HUD REGULATIONS REQUIRING NO IMPACT STATEMENT WITH RESPECT TO INITIAL HOUSING ACT GRANT APPLICATIONS CONTRAVENES NEPA. UNDER SECTION 104 OF THE HOUSING ACT, THE SECRETARY OF HUD, "IN LIEU OF THE ENVIRONMENTAL PROTECTION PROCEDURES OTHERWISE APPLICABLE, MAY UNDER REGULATIONS PROVIDE FOR THE RELEASE OF FUNDS FOR PARTICULAR PROJECTS TO APPLICANTS WHO ASSUME ALL OF THE RESPONSIBILITIES FOR ENVIRONEMTTAL REVIEW." HUD REGUALTIONS SAY THE PROCEDURE "ELIMINATES THE NECESSITY FOR FEDERAL ENVIRONMENTAL IMPACT STATEMENTS AT THE TIME OF THE INITIAL APPLICATION." JUDGE HENRY F. WERKER OF THE U.S. DISTRICT COURT FOR THE SOUTHERN DISTRICT OF NEW YORK SAID, "THIS IS PLAIN LANGUAGE, AND IN THE OPINION OF THIS COURT THE DEFENDANTS HAVE COMPLIED WITH THE RULE." CONGRESS "SPECIFICALLY AUTHORIZED THE SECRETARY OF HUD TO PROMULGATE REGULATIONS TO REPLACE NEPA'S ENVIRONMENTAL REVIEW REQUIREMENTS WHICH WOULD OTHERWISE BE APPLICABLE," THE COURT SAID, NOTING ALSO THAT THE CITY'S USE OF CITIZEN PARTICIPATION AND THE PLAN'S LONG- AND SHORT-TERM OBJECTIVES COMPLY WITH THE HOUSING ACT.
REFERENCE	ENVIRONMENT REPORTER, VOL. 6, NO. 29, NOV. 14, 1975, P. 1255

CASE	UNITED STATES VS. STOECO HOMES, INC., NO. 1335-T2
LOCATION	
DATE & STATUS	REINSTATED MARCH 14, 1974
PROJECT TYPE	CONSTRUCTION OF WATERFRONT HOMESITES
ISSUE(S)	WATER QUALITY, BUILDING PERMITS
SCOPE	THE COURT RULED THAT PRIVATE DREDGING AND CONSTRUCTION OF WATERFRONT HOMESITES UPON MARSHLAND INLAND OF RECOGNIZED HARBOR LINES AND NOT PRESENTLY WITHIN THE NAVIGABLE WATERWAYS OF THE UNITED STATES IS NOT SUBJECT TO FEDERAL PERMIT REQUIREMENTS UNDER THE RIVERS AND HARBORS APPROPRIATIONS ACT. THE ARMY CORPS OF ENGINEERS MAY ONLY ENJOIN SUCH ACTIVITY TO THE EXTENT THAT ACTUAL SILT DISCHARGE RESULTS FROM HYDRAULIC DREDGING. THE ARMY CLAIMED THAT THE LAND IS WITHIN THE NAVIGABLE WATERS OF THE U.S. HOWEVER, JUDGE GIBBONS, WRITING FOR THE COURT, DECLARED THAT THIS WAS INCONSISTENT WITH THE FACT THAT IN 1951, THE ARMY RECOGNIZED HARBOR LINES SEAWARD OF THE LANDS IN QUESTION. SINCE THEN, FURTHERMORE, THE LAND THAT WAS ONCE SUBJECT TO TIDES, HAS BECOME FAST LAND AND HENCE IS NOT SUBJECT TO FEDERAL PERMIT REQUIREMENTS.
REFERENCE	ENVIRONMENTAL LAW REPORTER, VOL. 4, P 20390

CASE	WALSH VS. SPADACCIA, NO. 07182-72
LOCATION	YORKTOWN, NEW YORK
DATE & STATUS	ENJOINED ON APRIL 20, 1973
PROJECT TYPE	APARTMENT BUILDING
ISSUE	ZONING REGULATIONS, WATER POLLUTION, TRAFFIC AND SCHOOL CONGESTION
SCOPE	THE NEW YORK SUPREME COURT RULED THAT A TOWN BOARD MUST GIVE INDEPENDENT AND MEANINGFUL CONSIDERATION TO ENVIRONMENTAL CONCERNS AS REQUIRED BY LOCAL ZONING AND TOWN PLAN ORDINANCES BEFORE IT CAN APPROVE A NEW DEVELOPMENT PROPOSAL SUBMITTED BY A PRIVATE DEVELOPER. SUIT WAS BROUGHT BY A GROUP OF HOMEOWNERS IN THE VICINITY OF THE PROPOSED SITE ON LAKE MOHEGAN OF THE 168 APARTMENT UNITS. IN RULING AGAINST THE DEFENDANTS, SUPREME COURT JUSTICE BEISHEIM NOTED THAT THERE ARE NO PUBLIC OR PRIVATE SEWERS IN THE AREA OF THE PROPOSED SITE. IN ISSUING PERMITS FOR THE CONSTRUCTION OF A SEPTIC TANK SYSTEM, THE TOWN BOARD HAD NOT MADE ANY STUDIES OF THE CONDITION OF LAKE MOHEGAN, NOR HAD ANYONE FROM THE COUNTY HEALTH DEPARTMENT PERSONALLY INSPECTED THE SITE. IN AT LEAST TWO INSTANCES THE PERMIT WAS IN VIOLATION OF THE REQUIRED DISTANCE BETWEEN THE SEPTIC FIELDS AND THE MARSH AREA ADJOINING THE LAKE. FURTHERMORE, THE COURT SAID, THE TOWN OF YORKTOWN REQUIRES APARTMENT HOUSES TO BE SERVICED BY PUBLIC SEWERS. "EVEN IF THE SEPTIC SYSTEM FOR THE PROPOSED CHALET HOME APARTMENTS SHOULD BE DEEMED TO HAVE MET COUNTY AND STATE REQUIREMENTS, THIS WOULD NOT ABROGATE THE NEED FOR COMPLIANCE WITH THE MORE STRINGENT TOWN REQUIREMENTS", THE COURT SAID. THE COURT RULED THAT THE TOWN BOARD ACTED IN AN ARBITRARY AND CAPRICIOUS MANNER IN NOT GIVING INDEPENDENT AND MEANINGFUL CONSIDERATION TO THE ISSUE OF SEWAGE DISPOSAL AND OF THE EFFECT OF THE APARTMENTS ON THE POLLUTION OF LAKE MOHEGAN. HOWEVER, THE COURT ALSO RULED THAT THE TOWN BOARD DID NOT ACT ARBITRARILY AND CAPRICIOUSLY IN APPROVING THE SITE PLAN AND DID GIVE PROPER CONSIDERATION TO THE IMPACT OF THE PROJECT ON TRAFFIC CONGESTION IN THE AREA AND THE IMPACT ON LOCAL SCHOOLS.
REFERENCE	ENVIRONMENTAL LAW REPORTER, VOL. 3 PP 20519-20522

CASE	WAMP VS. CHATTANOOGA HOUSING AUTHORITY, NO. 75-1192
LOCATION	CHATTANOOGA, TENNESSEE
DATE & STATUS	REINSTATED ON 5 DECEMBER, 1975
PROJECT TYPE	APARTMENT COMPLEX
ISSUE(S)	NEPA VIOLATION, JURISDICTION OF STATE AND DISTRICT COURT, ZONING CODES
SCOPE	THE U.S. COURT OF APPEALS FOR THE SIXTH CIRCUIT UPHELD A FEDERAL DISTRICT COURT'S DISMISSAL, FOR LACK OF STANDING, OF A CHALLENGE TO CONSTRUCTION OF AN APARTMENT COMPLEX. THE SUIT WAS FILED ORIGINALLY IN THE TENNESSEE CHANCERY COURT BY SEVERAL CHATTANOOGA TAXPAYERS WHO CHARGED THAT THE CHATTANOOGA HOUSING AUTHORITY AND HUD HAD ACTED IMPROPERLY IN APPROVING CONSTRUCTION OF THE PROJECT ON A FORMER PARKSITE. THE CASE WAS REMOVED BY THE DEFENDANTS TO THE U.S. DISTRICT COURT FOR THE EASTERN DISTRICT OF TENNESSEE, WHERE THE PLAINTIFFS SOUGHT TO AMEND THEIR COMPLAINT TO CHARGE THAT HUD, IN APPROVING THE SALE OF THE SITE TO PRIVATE DEVELOPERS, HAD VIOLATED THE NEPA. THE DISTRICT COURT DISMISSED THE COMPLAINT, RULING THAT TENNESSEE LAW REQUIRES PLAINTIFFS TO ESTABLISH THAT THEY WILL SUFFER "SOME SPECIAL INJURY NOT COMMON TO CITIZENS AND TAXPAYERS GENERALLY". BECAUSE PLAINTIFFS FAILED TO ESTABLISH SUCH A SPECIAL INJURY, THE DISTRICT COURT RULED, THEY LACKED STANDING IN THE STATE COURT, AND THE FEDERAL COURT THUS LACKED REMOVAL JURISDICTION. THE DISTRICT COURT "CORRECTLY CONSTRUED AND APPLIED THE CONTROLLING DECISIONS OF THE SUPREME COURT IN TENNESSEE" IN DETERMINING STATE STANDING REQUIREMENTS, THE APPEALS COURT RULED IN A PER CURIAM DECISION. "WE AGREE WITH THE DISTRICT COURT THAT IF APPELLANTS HAD NO STANDING TO MAINTAIN THE ACTION IN THE STATE COURT, THE DISTRICT COURT HAD NO REMOVAL JURISTICTION," THE COURT SAID.
REFERENCE	ENVIRONMENT REPORTER, PUBLISHED BY THE BUREAU OF NATIONAL AFFAIRS, INC., VOL. 6, NO. 34, DEC. 19, 1975, PP. 1411-1412 FOR THE FULL TEXT OF THE DISTRICT COURT'S OPINION, SEE: ENVIRONMENT REPORTER CASES, VOL. 7, P. 1250

CASE	WEST AUGUSTA HISTORICAL AND GENEALOGICAL SOCIETY VS. URBAN RENEWAL AUTHORITY OF PARKERSBURG, INC., NO. 73-2062
LOCATION	WEST VIRGINIA
DATE & STATUS	REMANDED JANUARY 22, 1974
PROJECT TYPE	DEMOLITION OF COUNTY COURTHOUSE, URBAN RENEWAL
ISSUE(S)	DESTRUCTION OF BUILDINGS WITH HISTORICAL SIGNIFICANCE (NHPA)
SCOPE	RULING ON A MOTION FOR DISMISSAL UNDER FEDERAL RULE OF CIVIL PROCEDURE, THE DISTRICT COURT HELD THAT THE OPPONENTS LACKED STANDING TO BRING AN ACTION TO ENJOIN THE DEMOLITION OF THE WOOD COUNTY COURTHOUSE AS A PART OF AN URBAN RENEWAL PROGRAM UNTIL HUD FILED AN EIS. ON APPEAL THE HISTORICAL SOCIETY PRESSED ITS RIGHT TO AMEND. LEAVE TO AMEND, ACCORDING TO THE FEDERAL RULE OF CIVIL PROCEDURE 15(A) "SHALL BE FREELY GIVEN WHEN JUSTICE SO REQUIRES". BECAUSE THE DEFENSE VOICED NO OBJECTION, THE COURT RULED THAT THE PROPOSED AMENDMENT WOULD CLARIFY THE HISTORICAL SOCIETY'S JUSTICIABLE INTEREST IN THIS LITIGATION. THUS, THE CIRCUIT COURT VACATED THE ORDER OF DISMISSAL AND REMANDED THE CASE TO THE DISTRICT COURT.
REFERENCE	ENVIRONMENTAL LAW REPORTER, VOL. 4, P 20229

CASE	WILSON VS. LYNN, NO. 74-392-S
LOCATION	BOSTON, MASSACHUSETTS
DATE & STATUS	REINSTATED MARCH 22, 1974
PROJECT TYPE	BUILDING REHABILITATION PROJECT, A PART OF URBAN RENEWAL PROJECT
ISSUE(S)	NEPA REQUIREMENTS, ALTERING PLANS AFTER EIS APPROVAL
SCOPE	THE COURT DENIED A MOTION TO ENJOIN A BUILDING REHABILITATION PROJECT. THE COURT'S DECISION WAS REACHED BECAUSE THE PLAINTIFFS WERE UNABLE TO PROVE THE PROJECT'S PROBABLE SIGNIFICANT EFFECT ON THE QUALITY OF THE HUMAN ENVIRONMENT WITHIN THE AREA. THE PROJECT AIMED AT REHABILITATING 36 BUILDINGS SO AS TO PRODUCE 185 DWELLING UNITS. THE PLAINTIFFS ARGUED THAT IN CONNECTION WITH HUD, HUD HAD NOT FILED THE EIS AS REQUIRED BY THE NEPA. HUD UNDERTOOK THE MORTGAGE INSURANCE AND THE GUARANTEE OF URBAN RENEWAL DEVELOPER'S INTEREST PAYMENTS. THE COURT FOUND HUD'S ACTION TO CONSTITUTE MAJOR FEDERAL ACTION. HOWEVER, THE COURT RULED THAT THE PLAINTIFFS' ASSERTION THAT THE RENEWAL PLAN WAS CHANGED AFTER ITS APPROVAL BY CITY COUNCIL, IS BASED ON NO EVIDENCE AND THEREFORE REQUIRED A NEW IMPACT STATEMENT. THUS THE COURT DENIED INJUNCTION RELIEF.
REFERENCE	ENVIRONMENTAL LAW REPORTER, VOL. 4, P 20476

FEDERAL ACTS

1. The National Environmental Policy Act of 1969, as amended

2. Council on Environmental Quality - Preparation of Environmental Impact Statements - Guidelines

1. The National Environmental Policy Act of 1969, as amended

An Act to establish a national policy for the environment, to provide for the establishment of a Council on Environmental Quality, and for other purposes.

Be it enacted by the Senate and House of Representatives of the United States of America in Congress assembled, That this Act may be cited as the "National Environmental Policy Act of 1969."

PURPOSE

SEC. 2. The purposes of this Act are: To declare a national policy which will encourage productive and enjoyable harmony between man and his environment; to promote efforts which will prevent or eliminate damage to the environment and biosphere and stimulate the health and welfare of man; to enrich the understanding of the ecological systems and natural resources important to the Nation; and to establish a Council on Environmental Quality.

TITLE I

DECLARATION OF NATIONAL ENVIRONMENTAL POLICY

SEC. 101. (a) The Congress, recognizing the profound impact of man's activity on the interrelations of all components of the natural environment, particularly the profound influences of population growth, high-density urbanization, industrial expansion, resource exploitation, and new and expanding technological advances and recognizing further the critical importance of restoring and maintaining environmental quality to the overall welfare and development of man, declares that it is the continuing policy of the Federal Government, in cooperation with State and local governments, and other concerned public and private organizations, to use all practicable means and measures, including financial and technical assistance, in a manner calculated to foster and promote the general welfare, to create and maintain conditions under which man and nature can exist in productive harmony, and fulfill the social, economic, and other requirements of present and future generations of Americans.

(b) In order to carry out the policy set forth in this Act, it is the continuing responsibility of the Federal Government to use all practicable means, consistent with other essential considerations of national policy, to improve

---

*Pub. L. 91–190, 42 U.S.C. 4321–4347, January 1, 1970, as amended by Pub. L. 94–83, August 9, 1975.

and coordinate Federal plans, functions, programs, and resources to the end that the Nation may—

(1) Fulfill the responsibilities of each generation as trustee of the environment for succeeding generations;

(2) Assure for all Americans safe, healthful, productive, and esthetically and culturally pleasuring surroundings;

(3) Attain the widest range of beneficial uses of the environment without degradation, risk to health or safety, or other undesirable and unintended consequences;

(4) Preserve important historic, cultural, and natural aspects of our national heritage, and maintain, wherever possible, an environment which supports diversity, and variety of individual choice;

(5) Achieve a balance between population and resource use which will permit high standards of living and a wide sharing of life's amenities; and

(6) Enhance the quality of renewable resources and approach the maximum attainable recycling of depletable resources.

(c) The Congress recognizes that each person should enjoy a healthful environment and that each person has a responsibility to contribute to the preservation and enhancement of the environment.

SEC. 102. The Congress authorizes and directs that, to the fullest extent possible: (1) the policies, regulations, and public laws of the United States shall be interpreted and administered in accordance with the policies set forth in this Act, and (2) all agencies of the Federal Government shall—

(A) Utilize a systematic, interdisciplinary approach which will insure the integrated use of the natural and social sciences and the environmental design arts in planning and in decisionmaking which may have an impact on man's environment;

(B) Identify and develop methods and procedures, in consultation with the Council on Environmental Quality established by title II of this Act, which will insure that presently unquantified environmental amenities and values may be given appropirate consideration in decisionmaking along with economic and technical considerations;

(C) Include in every recommendation or report on proposals for legislation and other major Federal actions significantly affecting the quality of the human environment, a detailed statement by the responsible official on—

(i) The environmental impact of the proposed action,

(ii) Any adverse environmental effects which cannot be avoided should the proposal be implemented,

(iii) Alternatives to the proposed action,

(iv) The relationship between local short-term uses of man's environment and the maintenance and enhancement of long-term productivity, and

(v) Any irreversible and irretrievable commitments of resources which would be involved in the proposed action should it be implemented.

Prior to making any detailed statement, the responsible Federal official shall consult with and obtain the comments of any Federal agency which has jurisdiction by law or special expertise with respect to any environmental impact involved. Copies of such statement and the comments and views of the appropriate Federal, State, and local agencies, which are authorized to develop and enforce environmental standards, shall be made available to the President, the Council on Environmental Quality and to the public as provided by section 552 of title 5, United States Code, and shall accompany the proposal through the existing agency review processes;

(D) Any detailed statement required under subparagraph (C) after January 1, 1970, for any major Federal action funded under a program of grants to States shall not be deemed to be legally insufficient solely by reason of having been prepared by a State agency or official, if:

(i) the State agency or official has statewide jurisdiction and has the responsibility for such action,

(ii) the responsible Federal official furnishes guidance and participates in such preparation,

(iii) the responsible Federal official independently evaluates such statement prior to its approval and adoption, and

(iv) after January 1, 1976, the responsible Federal official provides early notification to, and solicits the views of, any other State or any Federal land management entity of any action or any alternative thereto which may have significant impacts upon such State or affected Federal land management entity and, if there is any disagreement on such impacts, prepares a written assessment of such impacts and views for incorporation into such detailed statement.

The procedures in this subparagraph shall not relieve the Federal official of his responsibilities for the scope, objectivity, and content of the entire statement or of any other responsibility under this Act; and further, this subparagraph does not affect the legal sufficiency of statements prepared by State agencies with less than statewide jurisdiction.

(E) Study, develop, and describe appropriate alternatives to recommended courses of action in any proposal which involves unresolved conflicts concerning alternative uses of available resources;

(F) Recognize the worldwide and long-range character of environmental problems and, where consistent with the foreign policy of the United States, lend appropriate support to initiatives, resolutions, and programs designed to maximize international cooperation in anticipating and preventing a decline in the quality of mankind's world environment;

(G) Make available to States, counties, municipalities, institutions, and individuals, advice and information useful in restoring, maintaining, and enhancing the quality of the environment;

(H) Initiate and utilize ecological information in the planning and development of resource-oriented projects; and

(I) Assist the Council on Environmental Quality established by title II of this Act.

Sec. 103. All agencies of the Federal Government shall review their present statutory authority, administrative regulations, and current policies and procedures for the purpose of determining whether there are any deficiencies or inconsistencies therein which prohibit full compliance with the purposes and provisions of this Act and shall propose to the President not later than July 1, 1971, such measures as may be necessary to bring their authority and policies into conformity with the intent, purposes, and procedures set forth in this Act.

Sec. 104. Nothing in section 102 or 103 shall in any way affect the specific statutory obligations of any Federal agency (1) to comply with criteria or standards of environmental quality, (2) to coordinate or consult with any other Federal or State agency, or (3) to act, or refrain from acting contingent upon the recommendations or certification of any other Federal or State agency.

Sec. 105. The policies and goals set forth in this Act are supplementary to those set forth in existing authorizations of Federal agencies.

TITLE II

COUNCIL ON ENVIRONMENTAL QUALITY

Sec. 201. The President shall transmit to the Congress annually beginning July 1, 1970, an Environmental Quality Report (hereinafter referred to as the "report") which shall set forth (1) the status and condition of the major natural, manmade, or altered environmental classes of the Nation, including, but not limited to, the air, the aquatic, including marine, estuarine, and fresh water, and the terrestrial environment, including, but not limited to, the forest, dryland, wetland, range, urban, suburban and rural environment; (2) current and foreseeable trends in the quality, management and utilization of such environments and the effects of those trends on the social, economic, and other requirements of the Nation; (3) the adequacy of available natural resources for fulfilling human and economic requirements of the Nation in the light of expected population pressures; (4) a review of the programs and activities (including regulatory activities) of the Federal Government, the State and local governments, and nongovernmental entities or individuals with particular reference to their effect on the environment and on the conservation, development and utilization of natural resources; and (5) a program for remedying the deficiencies of existing programs and activities, together with recommendations for legislation.

SEC. 202. There is created in the Executive Office of the President a Council on Environmental Quality (hereinafter referred to as the "Council"). The Council shall be composed of three members who shall be appointed by the President to serve at his pleasure, by and with the advice and consent of the Senate. The President shall designate one of the members of the Council to serve as Chairman. Each member shall be a person who, as a result of his training, experience, and attainments, is exceptionally well qualified to analyze and interpret environmental trends and information of all kinds; to appraise programs and activities of the Federal Government in the light of the policy set forth in title I of this Act; to be conscious of and responsive to the scientific, economic, social, esthetic, and cultural needs and interests of the Nation; and to formulate and recommend national policies to promote the improvement of the quality of the environment.

SEC. 203. The Council may employ such officers and employees as may be necessary to carry out its functions under this Act. In addition, the Council may employ and fix the compensation of such experts and consultants as may be necessary for the carrying out of its functions under this Act, in accordance with section 3109 of title 5, United States Code (but without regard to the last sentence thereof).

SEC. 204. It shall be the duty and function of the Council—

(1) To assist and advise the President in the preparation of the Environmental Quality Report required by section 201;

(2) To gather timely and authoritative information concerning the conditions and trends in the quality of the environment both current and prospective, to analyze and interpret such information for the purpose of determining whether such conditions and trends are interfering, or are likely to interfere, with the achievement of the policy set forth in title I of this Act, and to compile and submit to the President studies relating to such conditions and trends;

(3) To review and appraise the various programs and activities of the Federal Government in the light of the policy set forth in title I of this Act for the purpose of determining the extent to which such programs and activities are contributing to the achievement of such policy, and to make recommendations to the President with respect thereto;

(4) To develop and recommend to the President national policies to foster and promote the improvement of environmental quality to meet the conservation, social, economic, health, and other requirements and goals of the Nation;

(5) To conduct investigations, studies, surveys, research, and analyses relating to ecological systems and environmental quality;

(6) To document and define changes in the natural environment, including the plant and animal systems, and to accumulate necessary

data and other information for a continuing analysis of these changes or trends and an interpretation of their underlying causes;

(7) To report at least once each year to the President on the state and condition of the environment; and

(8) To make and furnish such studies, reports thereon, and recommendations with respect to matters of policy and legislation as the President may request.

SEC. 205. In exercising its powers, functions, and duties under this Act, the Council shall—

(1) Consult with the Citizens' Advisory Committee on Environmental Quality established by Executive Order No. 11472, dated May 29, 1969, and with such representatives of science, industry, agriculture, labor, conservation organizations, State and local governments and other groups, as it deems advisable; and

(2) Utilize, to the fullest extent possible, the services, facilities and information (including statistical information) of public and private agencies and organizations, and individuals, in order that duplication of effort and expense may be avoided, thus assuring that the Council's activities will not unnecessarily overlap or conflict with similar activities authorized by law and performed by established agencies.

SEC. 206. Members of the Council shall serve full time and the Chairman of the Council shall be compensated at the rate provided for Level II of the Executive Schedule Pay Rates (5 U.S.C. 5313). The other members of the Council shall be compensated at the rate provided for Level IV of the Executive Schedule Pay Rates (5 U.S.C. 5315).

SEC. 207. There are authorized to be appropriated to carry out the provisions of this Act not to exceed $300,000 for fiscal year 1970, $700,000 for fiscal year 1971, and $1 million for each fiscal year thereafter.

Approved January 1, 1970.

2. COUNCIL ON ENVIRONMENTAL QUALITY - PREPARATION OF ENVIRONMENTAL IMPACT STATEMENTS - Guidelines

federal register

WEDNESDAY, AUGUST 1, 1973
WASHINGTON, D.C.

Volume 38 ■ Number 147

PART II

COUNCIL ON ENVIRONMENTAL QUALITY

■

PREPARATION OF ENVIRONMENTAL IMPACT STATEMENTS

Guidelines

Title 40—Protection of the Environment

CHAPTER V—COUNCIL ON ENVIRONMENTAL QUALITY

PART 1500—PREPARATION OF ENVIRONMENTAL IMPACT STATEMENTS: GUIDELINES

On May 2, 1973, the Council on Environmental Quality published in the FEDERAL REGISTER, for public comment, a proposed revision of its guidelines for the preparation of environmental impact statements. Pursuant to the National Environmental Policy Act (P.L. 91–190, 42 U.S.C. 4321 et seq.) and Executive Order 11514 (35 FR 4247) all Federal departments, agencies, and establishments are required to prepare such statements in connection with their proposals for legislation and other major Federal actions significantly affecting the quality of the human environment. The authority for the Council's guidelines is set forth below in § 1500.1. The specific policies to be implemented by the guidelines is set forth below in § 1500.2.

The Council received numerous comments on its proposed guidelines from environmental groups, Federal, State, and local agencies, industry, and private individuals. Two general themes were presented in the majority of the comments. First, the Council should increase the opportunity for public involvement in the impact statement process. Second, the Council should provide more detailed guidance on the responsibilities of Federal agencies in light of recent court decisions interpreting the Act. The proposed guidelines have been revised in light of the specific comments relating to these general themes, as well as other comments received, and are now being issued in final form.

The guidelines will appear in the Code of Federal Regulations in Title 40, Chapter V, at Part 1500. They are being codified, in part, because they affect State and local governmental agencies, environmental groups, industry, and private individuals, in addition to Federal agencies, to which they are specifically directed, and the resultant need to make them widely and readily available.

Sec.

1500.1 Purpose and authority.
1500.2 Policy.
1500.3 Agency and OMB procedures.
1500.4 Federal agencies included; effect of the act on existing agency mandates.
1500.5 Types of actions covered by the act.
1500.6 Identifying major actions significantly affecting the environment.
1500.7 Preparing draft environmental statements; public hearings.
1500.8 Content of environmental statements.
1500.9 Review of draft environmental statements by Federal, Federal-State, and local agencies and by the public.
1500.10 Preparation and circulation of final environmental statements.
1500.11 Transmittal of statements to the Council; minimum periods for review; requests by the Council.
1500.12 Legislative actions.
1500.13 Application of section 102(2)(C) procedure to existing projects and programs.
1500.14 Supplementary guidelines; evaluation of procedures.

Sec.

Appendix I Summary to accompany draft and final statements.

Appendix II Areas of environmental impact and Federal agencies and Federal State agencies with jurisdiction by law or special expertise to comment thereon.

Appendix III Offices within Federal agencies and Federal-State agencies for information regarding the agencies' NEPA activities and for receiving other agencies' impact statements for which comments are requested.

Appendix IV State and local agency review of impact statements.

AUTHORITY: National Environmental Act (P.L. 91–190, 42 U.S.C. 4321 et seq.) and Executive Order 11514.

§ 1500.1 Purpose and authority.

(a) This directive provides guidelines to Federal departments, agencies, and establishments for preparing detailed environmental statements on proposals for legislation and other major Federal actions significantly affecting the quality of the human environment as required by section 102(2)(C) of the National Environmental Policy Act (P.L. 91–190, 42 U.S.C. 4321 et. seq.) (hereafter "the Act"). Underlying the preparation of such environmental statements is the mandate of both the Act and Executive Order 11514 (35 FR 4247) of March 5, 1970, that all Federal agencies, to the fullest extent possible, direct their policies, plans and programs to protect and enhance environmental quality. Agencies are required to view their actions in a manner calculated to encourage productive and enjoyable harmony between man and his environment, to promote efforts preventing or eliminating damage to the environment and biosphere and stimulating the health and welfare of man, and to enrich the understanding of the ecological systems and natural resources important to the Nation. The objective of section 102(2)(C) of the Act and of these guidelines is to assist agencies in implementing these policies. This requires agencies to build into their decisionmaking process, beginning at the earliest possible point, an appropriate and careful consideration of the environmental aspects of proposed action in order that adverse environmental effects may be avoided or minimized and environmental quality previously lost may be restored. This directive also provides guidance to Federal, State, and local agencies and the public in commenting on statements prepared under these guidelines.

(b) Pursuant to section 204(3) of the Act the Council on Environmental Quality (hereafter "the Council") is assigned the duty and function of reviewing and appraising the programs and activities of the Federal Government, in the light of the Act's policy, for the purpose of determining the extent to which such programs and activities are contributing to the achievement of such policy, and to make recommendations to the President with respect thereto. Section 102(2)(B) of the Act directs all Federal agencies to identify and develop methods and procedures, in consultation with the Council, to insure that unquantified environmental values be given appropriate consideration in decisionmaking along with economic and technical considerations; section 102(2)(C) of the Act directs that copies of all environmental impact statements be filed with the Council; and section 102(2)(H) directs all Federal agencies to assist the Council in the performance of its functions. These provisions have been supplemented in sections 3(h) and (i) of Executive Order 11514 by directions that the Council issue guidelines to Federal agencies for preparation of environmental impact statements and such other instructions to agencies and requests for reports and information as may be required to carry out the Council's responsibilities under the Act.

§ 1500.2 Policy.

(a) As early as possible and in all cases prior to agency decision concerning recommendations or favorable reports on proposals for (1) legislation significantly affecting the quality of the human environment (see §§ 1500.5(i) and 1500.12) (hereafter "legislative actions") and (2) all other major Federal actions significantly affecting the quality of the human environment (hereafter "administrative actions"), Federal agencies will, in consultation with other appropriate Federal, State and local agencies and the public assess in detail the potential environmental impact.

(b) Initial assessments of the environmental impacts of proposed action should be undertaken concurrently with initial technical and economic studies and, where required, a draft environmental impact statement prepared and circulated for comment in time to accompany the proposal through the existing agency review processes for such action. In this process, Federal agencies shall: (1) Provide for circulation of draft environmental statements to other Federal, State, and local agencies and for their availability to the public in accordance with the provisions of these guidelines; (2) consider the comments of the agencies and the public; and (3) issue final environmental impact statements responsive to the comments received. The purpose of this assessment and consultation process is to provide agencies and other decisionmakers as well as members of the public with an understanding of the potential environmental effects of proposed actions, to avoid or minimize adverse effects wherever possible, and to restore or enhance environmental quality to the fullest extent practicable. In particular, agencies should use the environmental impact statement process to explore alternative actions that will avoid or minimize adverse impacts and to evaluate both the long- and short-range implications of proposed actions to man, his physical and social surroundings, and to nature. Agencies should consider the results of their environmental assessments along with their assessments of the net economic, technical and other benefits of proposed actions and use all practicable means, consistent with other essential considerations of national policy, to restore environmental quality as well as to avoid or minimize undesirable consequences for the environment.

§ 1500.3 **Agency and OMB procedures.**

(a) Pursuant to section 2(f) of Executive Order 11514, the heads of Federal agencies have been directed to proceed with measures required by section 102 (2)(C) of the Act. Previous guidelines of the Council directed each agency to establish its own formal procedures for (1) identifying those agency actions requiring environmental statements, the appropriate time prior to decision for the consultations required by section 102 (2)(C) and the agency review process for which environmental statements are to be available, (2) obtaining information required in their preparation, (3) designating the officials who are to be responsible for the statements, (4) consulting with and taking account of the comments of appropriate Federal, State and local agencies and the public, including obtaining the comment of the Administrator of the Environmental Protection Agency when required under section 309 of the Clean Air Act, as amended, and (5) meeting the requirements of section 2(b) of Executive Order 11514 for providing timely public information on Federal plans and programs with environmental impact. Each agency, including both departmental and subdepartmental components having such procedures, shall review its procedures and shall revise them, in consultation with the Council, as may be necessary in order to respond to requirements imposed by these revised guidelines as well as by such previous directives. After such consultation, proposed revisions of such agency procedures shall be published in the FEDERAL REGISTER no later than October 30, 1973. A minimum 45-day period for public comment shall be provided, followed by publication of final procedures no later than forty-five (45) days after the conclusion of the comment period. Each agency shall submit seven (7) copies of all such procedures to the Council. Any future revision of such agency procedures shall similarly be proposed and adopted only after prior consultation with the Council and, in the case of substantial revision, opportunity for public comment. All revisions shall be published in the FEDERAL REGISTER.

(b) Each Federal agency should consult, with the assistance of the Council and the Office of Management and Budget if desired, with other appropriate Federal agencies in the development and revision of the above procedures so as to achieve consistency in dealing with similar activities and to assure effective coordination among agencies in their review of proposed activities. Where applicable, State and local review of such agency procedures should be conducted pursuant to procedures established by Office of Management and Budget Circular No. A-85.

(c) Existing mechanisms for obtaining the views of Federal, State, and local agencies on proposed Federal actions should be utilized to the maximum extent practicable in dealing with environmental matters. The Office of Management and Budget will issue instructions, as necessary, to take full advantage of such existing mechanisms.

§ 1500.4 **Federal agencies included; effect of the Act on existing agency mandates.**

(a) Section 102(2)(C) of the Act applies to all agencies of the Federal Government. Section 102 of the Act provides that "to the fullest extent possible: (1) The policies, regulations, and public laws of the United States shall be interpreted and administered in accordance with the policies set forth in this Act," and section 105 of the Act provides that "the policies and goals set forth in this Act are supplementary to those set forth in existing authorizations of Federal agencies." This means that each agency shall interpret the provisions of the Act as a supplement to its existing authority and as a mandate to view traditional policies and missions in the light of the Act's national environmental objectives. In accordance with this purpose, agencies should continue to review their policies, procedures, and regulations and to revise them as necessary to ensure full compliance with the purposes and provisions of the Act. The phrase "to the fullest extent possible" in section 102 is meant to make clear that each agency of the Federal Government shall comply with that section unless existing law applicable to the agency's operations expressly prohibits or makes compliance impossible.

§ 1500.5 **Types of actions covered by the Act.**

(a) "Actions" include but are not limited to:

(1) Recommendations or favorable reports relating to legislation including requests for appropriations. The requirement for following the section 102 (2)(C) procedure as elaborated in these guidelines applies to both (i) agency recommendations on their own proposals for legislation (see § 1500.12); and (ii) agency reports on legislation initiated elsewhere. In the latter case only the agency which has primary responsibility for the subject matter involved will prepare an environmental statement.

(2) New and continuing projects and program activities: directly undertaken by Federal agencies; or supported in whole or in part through Federal contracts, grants, subsidies, loans, or other forms of funding assistance (except where such assistance is solely in the form of general revenue sharing funds, distributed under the State and Local Fiscal Assistance Act of 1972, 31 U.S.C. 1221 et. seq. with no Federal agency control over the subsequent use of such funds); or involving a Federal lease, permit, license certificate or other entitlement for use.

(3) The making, modification, or establishment of regulations, rules, procedures, and policy.

§ 1500.6 **Identifying major actions significantly affecting the environment.**

(a) The statutory clause "major Federal actions significantly affecting the quality of the human environment" is to be construed by agencies with a view to the overall, cumulative impact of the action proposed, related Federal actions and projects in the area, and further actions contemplated. Such actions may be localized in their impact, but if there is potential that the environment may be significantly affected, the statement is to be prepared. Proposed major actions, the enviornmental impact of which is likely to be highly controversial, should be covered in all cases. In considering what constitutes major action significantly affecting the environment, agencies should bear in mind that the effect of many Federal decisions about a project or complex of projects can be individually limited but cumulatively considerable. This can occur when one or more agencies over a period of years puts into a project individually minor but collectively major resources, when one decision involving a limited amount of money is a precedent for action in much larger cases or represents a decision in principle about a future major course of action, or when several Government agencies individually make decisions about partial aspects of a major action. In all such cases, an environmental statement should be prepared if it is reasonable to anticipate a cumulatively significant impact on the environment from Federal action. The Council, on the basis of a written assessment of the impacts involved, is available to assist agencies in determining whether specific actions require impact statements.

(b) Section 101(b) of the Act indicates the broad range of aspects of the environment to be surveyed in any assessment of significant effect. The Act also indicates that adverse significant effects include those that degrade the quality of the environment, curtail the range of beneficial uses of the environment, and serve short-term, to the disadvantage of long-term, environmental goals. Significant effects can also include actions which may have both beneficial and detrimental effects, even if on balance the agency believes that the effect will be beneficial. Significant effects also include secondary effects, as described more fully, for example, in § 1500.8(a) (iii)(B). The significance of a proposed action may also vary with the setting, with the result that an action that would have little impact in an urban area may be significant in a rural setting or vice versa. While a precise definition of environmental "significance," valid in all contexts, is not possible, effects to be considered in assessing significance include, but are not limited to, those outlined in Appendix II of these guidelines.

(c) Each of the provisions of the Act, except section 102(2)(C), applies to all Federal agency actions. Section 102(2) (C) requires the preparation of a detailed environmental impact statement in the case of "major Federal actions significantly affecting the quality of the human environment." The identification of major actions significantly affecting the environment is the responsibility of each Federal agency, to be carried out against the background of its own particular operations. The action must be a (1) "major" action, (2) which is a "Federal action," (3) which has a "significant" effect, and (4) which involves the "quality of the human environment." The words "major" and "significantly" are intended to imply thresholds of importance and impact that must be met before a statement is required. The action causing the impact must also be one where there is sufficient Federal control and responsibility to constitute "Federal action" in contrast to cases where such Federal control and responsibility are not present as, for example, when Federal funds are distributed in the form of general revenue sharing to be used by State and local governments (see § 1500.5(ii)). Finally, the action must be one that significantly affects the quality of the human environment either by directly affecting human beings or by indirectly affecting human beings through adverse effects on the environment. Each agency should review the typical classes of actions that it undertakes and, in consultation with the Council, should develop specific criteria and methods for identifying those actions likely to require environmental statements and those actions likely not to require environmental statements. Normally this will involve:

(i) Making an initial assessment of the environmental impacts typically associated with principal types of agency action.

(ii) Identifying on the basis of this assessment, types of actions which normally do, and types of actions which normally do not, require statements.

(iii) With respect to remaining actions that may require statements depending on the circumstances, and those actions determined under the preceding paragraph (C)(4)(ii) of this section as likely to require statements, identifying: (a) what basic information needs to be gathered; (b) how and when such information is to be assembled and analyzed; and (c) on what bases environmental assessments and decisions to prepare impact statements will be made. Agencies may either include this substantive guidance in the procedures issued pursuant to § 1500.3(a) of these guidelines, or issue such guidance as supplemental instructions to aid relevant agency personnel in implementing the impact statement process. Pursuant to § 1500.14 of these guidelines, agencies shall report to the Council by June 30, 1974, on the progress made in developing such substantive guidance.

(d) (1) Agencies should give careful attention to identifying and defining the purpose and scope of the action which would most appropriately serve as the subject of the statement. In many cases, broad program statements will be required in order to assess the environmental effects of a number of individual actions on a given geographical area (e.g., coal leases), or environmental impacts that are generic or common to a series of agency actions (e.g., maintenance or waste handling practices), or the overall impact of a large-scale program or chain of contemplated projects (e.g., major lengths of highway as opposed to small segments). Subsequent statements on major individual actions will be necessary where such actions have significant environmental impacts not adequately evaluated in the program statement.

(2) Agencies engaging in major technology research and development programs should develop procedures for periodic evaluation to determine when a program statement is required for such programs. Factors to be considered in making this determination include the magnitude of Federal investment in the program, the likelihood of widespread application of the technology, the degree of environmental impact which would occur if the technology were widely applied, and the extent to which continued investment in the new technology is likely to restrict future alternatives. Statements must be written late enough in the development process to contain meaningful information, but early enough so that this information can practically serve as an input in the decision-making process. Where it is anticipated that a statement may ultimately be required but that its preparation is still premature, the agency should prepare an evaluation briefly setting forth the reasons for its determination that a statement is not yet necessary. This evaluation should be periodically updated, particularly when significant new information becomes available concerning the potential environmental impact of the program. In any case, a statement must be prepared before research activities have reached a stage of investment or commitment to implementation likely to determine subsequent development or restrict later alternatives. Statements on technology research and development programs should include an analysis not only of alternative forms of the same technology that might reduce any adverse environmental impacts but also of alternative technologies that would serve the same function as the technology under consideration. Efforts should be made to involve other Federal agencies and interested groups with relevant expertise in the preparation of such statements because the impacts and alternatives to be considered are likely to be less well defined than in other types of statements.

(e) In accordance with the policy of the Act and Executive Order 11514 agencies have a responsibility to develop procedures to insure the fullest practicable provision of timely public information and understanding of Federal plans and programs with environmental impact in order to obtain the views of interested parties. In furtherance of this policy, agency procedures should include an appropriate early notice system for informing the public of the decision to prepare a draft environmental statement on proposed administrative actions (and for soliciting comments that may be helpful in preparing the statement) as soon as is practicable after the decision to prepare the statement is made. In this connection, agencies should: (1) maintain a list of administrative actions for which environmental statements are being prepared; (2) revise the list at regular intervals specified in the agency's procedures developed pursuant to § 1500.3(a) of these guidelines (but not less than quarterly) and transmit each such revision to the Council; and (3) make the list available for public inspection on request. The Council will periodically publish such lists in the FEDERAL REGISTER. If an agency decides that an environmental statement is not necessary for a proposed action (i) which the agency has identified pursuant to § 1500.6(c)(4)(ii) as normally requiring preparation of a statement, (ii) which is similar to actions for which the agency has prepared a significant number of statements, (iii) which the agency has previously announced would be the subject of a statement, or (iv) for which the agency has made a negative determination in response to a request from the Council pursuant to § 1500.11(f), the agency shall prepare a publicly available record briefly setting forth the agency's decision and the reasons for that determination. Lists of such negative determinations, and any evaluations made pursuant to § 1500.6 which conclude that preparation of a statement is not yet timely, shall be prepared and made available in the same manner as provided in this subsection for lists of statements under preparation.

§ 1500.7 Preparing draft environmental statements; public hearings.

(a) Each environmental impact statement shall be prepared and circulated in draft form for comment in accordance with the provisions of these guidelines. The draft statement must fulfill and satisfy to the fullest extent possible at the time the draft is prepared the requirements established for final statements by section 102(2)(C). (Where an agency has an established practice of declining to favor an alternative until public comments on a proposed action have been received, the draft environmental statement may indicate that two or more alternatives are under consideration.) Comments received shall be carefully evaluated and considered in the decision process. A final statement with substantive comments attached shall then be issued and circulated in accordance with applicable provisions of §§ 1500.10, 1500.11, or 1500.12. It is important that draft environmental statements be prepared and circulated for comment and furnished to the Council as early as possible in the agency review process in order to permit agency decisionmakers and outside reviewers to give meaningful consideration to the environmental issues involved. In particular, agencies should keep in mind that such statements are to serve as the means of assessing the environmental impact of proposed agency actions, rather than as a justification for decisions already made. This means that draft statements on administrative actions should be prepared and circulated for comment prior to the first significant point of decision in the agency review process. For major categories of agency action, this point should be identified in the procedures issued pursuant to § 1500.3(a). For major categories of projects involving an applicant and identified pursuant to § 1500.6 (c)(c)(ii) as normally requiring the preparation of a statement, agencies should include in their procedures provisions limiting actions which an applicant is permitted to take prior to completion and review of the final statement with respect to his application.

(b) Where more than one agency (1) directly sponsors an action, or is directly involved in an action through funding, licenses, or permits, or (2) is involved in a group of actions directly related to each other because of their functional interdependence and geographical proximity, consideration should be given to preparing one statement for all the Federal actions involved (see § 1500.6(d)(1)). Agencies in such cases should consider the possibility of joint preparation of a statement by all agencies concerned, or designation of a single "lead agency" to assume supervisory responsibility for preparation of the statement. Where a lead agency prepares the statement, the other agencies involved should provide assistance with respect to their areas of jurisdiction and expertise. In either case, the statement should contain an environmental assessment of the full range of Federal actions involved, should reflect the views of all participating agencies, and should be prepared before major or irreversible actions have been taken by any of the participating agencies. Factors relevant in determining an appropriate lead agency include the time sequence in which the agencies become involved, the magnitude of their respective involvement, and their relative expertise with respect to the project's environmental effects. As necessary, the Council will assist in resolving questions of responsibility for statement preparation in the case of multi-agency actions. Federal Regional Councils, agencies and the public are encouraged to bring to the attention of the Council and other relevant agencies appropriate situations where a geographic or regionally focused statement would be desirable because of the cumulative environmental effects likely to result from multi-agency actions in the area.

(c) Where an agency relies on an applicant to submit initial environmental information, the agency should assist the applicant by outlining the types of information required. In all cases, the agency should make its own evaluation of the environmental issues and take responsibility for the scope and content of draft and final environmental statements.

(d) Agency procedures developed pursuant to § 1500.3(a) of these guidelines should indicate as explicitly as possible those types of agency decisions or actions which utilize hearings as part of the normal agency review process, either as a result of statutory requirement or agency practice. To the fullest extent possible, all such hearings shall include consideration of the environmental aspects of the proposed action. Agency procedures shall also specifically include provision for public hearings on major actions with environmental impact, whenever appropriate, and for providing the public with relevant information, including information on alternative courses of action. In deciding whether a public hearing is appropriate, an agency should consider: (1) The magnitude of the proposal in terms of economic costs, the geographic area involved, and the uniqueness or size of commitment of the resources involved; (2) the degree of interest in the proposal, as evidenced by requests from the public and from Federal, State and local authorities that a hearing be held; (3) the complexity of the issue and the likelihood that information will be presented at the hearing which will be of assistance to the agency in fulfilling its responsibilities under the Act; and (4) the extent to which public involvement already has been achieved through other means, such as earlier public hearings, meetings with citizen representatives, and/or written comments on the proposed action. Agencies should make any draft environmental statements to be issued available to the public at least fifteen (15) days prior to the time of such hearings.

§ 1500.8 Content of environmental statements.

(a) The following points are to be covered:

(1) A description of the proposed action, a statement of its purposes, and a description of the environment affected, including information, summary technical data, and maps and diagrams where relevant, adequate to permit an assessment of potential environmental impact by commenting agencies and the public. Highly technical and specialized analyses and data should be avoided in the body of the draft impact statement. Such materials should be attached as appendices or footnoted with adequate bibliographic references. The statement should also succinctly describe the environment of the area affected as it exists prior to a proposed action, including other Federal activities in the area affected by the proposed action which are related to the proposed action. The interrelationships and cumulative environmental impacts of the proposed action and other related Federal projects shall be presented in the statement. The amount of detail provided in such descriptions should be commensurate with the extent and expected impact of the action, and with the amount of information required at the particular level of decisionmaking (planning, feasibility, design, etc.). In order to ensure accurate descriptions and environmental assessments, site visits should be made where feasible. Agencies should also take care to identify, as appropriate, population and growth characteristics of the affected area and any population and growth assumptions used to justify the project or program or to determine secondary population and growth impacts resulting from the proposed action and its alternatives (see paragraph (a)(1)(3)(ii), of this section). In discussing these population aspects, agencies should give consideration to using the rates of growth in the region of the project contained in the projection compiled for the Water Resources Council by the Bureau of Economic Analysis of the Department of Commerce and the Economic Research Service of the Department of Agriculture (the "OBERS" projection). In any event it is essential that the sources of data used to identify, quantify or evaluate any and all environmental consequences be expressly noted.

(2) The relationship of the proposed action to land use plans, policies, and controls for the affected area. This requires a discussion of how the proposed action may conform or conflict with the objectives and specific terms of approved or proposed Federal, State, and local land use plans, policies, and controls, if any, for the area affected including those developed in response to the Clean Air Act or the Federal Water Pollution Control Act Amendments of 1972. Where a conflict or inconsistency exists, the statement should describe the extent to which the agency has reconciled its proposed action with the plan, policy or control, and the reasons why the agency has decided to proceed notwithstanding the absence of full reconciliation.

(3) The probable impact of the proposed action on the environment.

(i) This requires agencies to assess the positive and negative effects of the proposed action as it affects both the national and international environment. The attention given to different environmental factors will vary according to the nature, scale, and location of proposed actions. Among factors to consider should be the potential effect of the action on such aspects of the environment as those listed in Appendix II of these guidelines. Primary attention should be given in the statement to discussing those factors most evidently impacted by the proposed action.

(ii) Secondary or indirect, as well as primary or direct, consequences for the environment should be included in the analysis. Many major Federal actions, in particular those that involve the construction or licensing of infrastructure investments (e.g., highways, airports, sewer systems, water resource projects, etc.), stimulate or induce secondary effects in the form of associated investments and changed patterns of social and economic activities. Such secondary effects, through their impacts on existing community facilities and activities, through inducing new facilities and activities, or through changes in natural conditions, may often be even more substantial than the primary effects of the original action itself. For example, the effects of the proposed action on population and growth may be among the more significant secondary effects. Such population and growth impacts should be estimated if expected to be significant (using data identified as indicated in § 1500.8(a)(1)) and an assessment made of the effect of any possible change in population patterns or growth upon the resource base, including land use, water, and public services, of the area in question.

(4) Alternatives to the proposed action, including, where relevant, those not within the existing authority of the responsible agency. (Section 102(2)(D) of the Act requires the responsible agency to "study, develop, and describe appropriate alternatives to recommended courses of action in any proposal which involves unresolved conflicts concerning alternative uses of available resources"). A rigorous exploration and objective evaluation of the environmental impacts of all reasonable alternative actions, particularly those that might enhance environmental quality or avoid some or all of the adverse environmental effects, is essential. Sufficient analysis of such alternatives and their environmental benefits, costs and risks should accompany the proposed action through the agency review process in order not to foreclose prematurely options which might enhance environmental quality or have less detrimental effects. Examples of such alternatives include: the alternative of taking no action or of postponing action pending further study; alternatives requiring actions of a significantly different nature which would provide similar benefits with different environmental impacts (e.g., nonstructural alternatives to flood control programs, or mass transit alternatives to highway construction); alternatives related to different designs or details of the proposed action which would present different environmental impacts (e.g., cooling ponds vs. cooling towers for a power plant or alternatives that will significantly conserve energy); alternative measures to provide for compensation of fish and wildlife losses, including the acquisition of land, waters, and interests therein. In each case, the analysis should be sufficiently detailed to reveal the agency's comparative evaluation of the environmental benefits, costs and risks of the proposed action and each reasonable alternative. Where an existing impact statement already contains such an analysis, its treatment of alternatives may be incorporated provided that such treatment is current and relevant to the precise purpose of the proposed action.

(5) Any probable adverse environmental effects which cannot be avoided (such as water or air pollution, undesirable land use patterns, damage to life systems, urban congestion, threats to health or other consequences adverse to the environmental goals set out in section 101(b) of the Act). This should be a brief section summarizing in one place those effects discussed in paragraph (a)(3) of this section that are adverse and unavoidable under the proposed action. Included for purposes of contrast should be a clear statement of how other avoidable adverse effects discussed in paragraph (a)(2) of this section will be mitigated.

(6) The relationship between local short-term uses of man's environment and the maintenance and enhancement of long-term productivity. This section should contain a brief discussion of the extent to which the proposed action involves tradeoffs between short-term environmental gains at the expense of long-term losses, or vice versa, and a discussion of the extent to which the proposed action forecloses future options. In this context short-term and long-term do not refer to any fixed time periods, but should be viewed in terms of the environmentally significant consequences of the proposed action.

(7) Any irreversible and irretrievable commitments of resources that would be involved in the proposed action should it be implemented. This requires the agency to identify from its survey of unavoidable impacts in paragraph (a)(5) of this section the extent to which the action irreversibly curtails the range of potential uses of the environment. Agencies should avoid construing the term "resources" to mean only the labor and materials devoted to an action. "Resources" also means the natural and cultural resources committed to loss or destruction by the action.

(8) An indication of what other interests and considerations of Federal policy are thought to offset the adverse environmental effects of the proposed action identified pursuant to paragraphs (a)(3) and (5) of this section. The statement should also indicate the extent to which these stated countervailing benefits could be realized by following reasonable alternatives to the proposed action (as identified in paragraph (a)(4) of this section) that would avoid some or all of the adverse environmental effects. In this connection, agencies that prepare cost-benefit analyses of proposed actions should attach such analyses, or summaries thereof, to the environmental impact statement, and should clearly indicate the extent to which environmental costs have not been reflected in such analyses.

(b) In developing the above points agencies should make every effort to convey the required information succinctly in a form easily understood, both by members of the public and by public decisionmakers, giving attention to the substance of the information conveyed rather than to the particular form, or length, or detail of the statement. Each of the above points, for example, need not always occupy a distinct section of the statement if it is otherwise adequately covered in discussing the impact of the proposed action and its alternatives—which items should normally be the focus of the statement. Draft statements should indicate at appropriate points in the text any underlying studies, reports, and other information obtained and considered by the agency in preparing the statement including any cost-benefit analyses prepared by the agency, and reports of consulting agencies under the Fish and Wildlife Coordination Act, 16 U.S.C 661 et seq., and the National Historic Preservation Act of 1966, 16 U.S.C. 470 et seq., where such consultation has taken place. In the case of documents not likely to be easily accessible (such as internal studies or reports), the agency should indicate how such information may be obtained. If such information is attached to the statement, care should be taken to ensure that the statement remains an essentially self-contained instrument, capable of being understood by the reader without the need for undue cross reference.

(c) Each environmental statement should be prepared in accordance with the precept in section 102(2)(A) of the Act that all agencies of the Federal Government "utilize a systematic, interdisciplinary approach which will insure the integrated use of the natural and social sciences and the environmental design arts in planning and decisionmaking which may have an impact on man's environment." Agencies should attempt to have relevant disciplines represented on their own staffs; where this is not feasible they should make appropriate use of relevant Federal, State, and local agencies or the professional services of universities and outside consultants. The interdisciplinary approach should not be limited to the preparation of the environmental impact statement, but should also be used in the early planning stages of the proposed action. Early application of such an approach should help assure a systematic evaluation of reasonable alternative courses of action and their potential social, economic, and environmental consequences.

(d) Appendix I prescribes the form of the summary sheet which should accompany each draft and final environmental statement.

§ 1500.9 Review of draft environmental statements by Federal, Federal-State, State, and local agencies and by the public.

(a) *Federal agency review.* (1) *In general.* A Federal agency considering an action requiring an environmental statement should consult with, and (on the basis of a draft environmental statement for which the agency takes responsibility) obtain the comment on the environmental impact of the action of Federal and Federal-State agencies with jurisdiction by law or special expertise with respect to any environmental impact involved. These Federal and Federal-State agencies and their relevant areas of expertise include those identified in Appendices II and III to these guidelines. It is recommended that the listed departments and agencies establish contact points, which may be regional offices, for providing comments on the environmental statements. The requirement in section 102(2)(C) to obtain comment from Federal agencies having jurisdiction or special expertise is in addition to any specific statutory obligation of any Federal agency to coordinate or consult with any other Federal or State agency. Agencies should, for example, be alert to consultation requirements of the Fish and Wildlife Coordination Act, 16 U.S.C. 661 et seq., and the National Historic Preservation Act of 1966, 16 U.S.C. 470 et seq. To the extent possible, statements or findings concerning environmental impact required by other statutes, such as section 4(f) of the Department of Transportation Act of 1966, 49 U.S.C. 1653(f), or section 106 of the National Historic Preservation Act of 1966, should be combined with compliance with the environmental impact statement requirements of section 102(2)(C) of the Act to yield a single document which meets all applicable requirements. The Advisory Council on Historic Preservation, the Department of Transportation,* and the Department of the Interior, in consultation with the Council, will issue any necessary supplementing instructions for furnishing information or findings not forthcoming under the environmental impact statement process.

(b) *EPA review.* Section 309 of the Clean Air Act, as amended (42 U.S.C. § 1857h–7), provides that the Administrator of the Environmental Protection Agency shall comment in writing on the environmental impact of any matter relating to his duties and responsibilities, and shall refer to the Council any matter that the Administrator determines is unsatisfactory from the standpoint of public health or welfare or environmental quality. Accordingly, wherever an agency action related to air or water quality, noise abatement and control, pesticide regulation, solid waste disposal, generally applicable environmental radiation criteria and standards, or other provision of the authority of the Administrator is involved, Federal agencies are required to submit such proposed actions and their environmental impact statements, if such have been prepared, to the Administrator for review and comment in writing. In all cases where EPA determines that proposed agency action is environmentally unsatisfactory, or where EPA determines that an environmental statement is so inadequate that such a determination cannot be made, EPA shall publish its determination and notify the Council as soon as practicable. The Administrator's comments shall constitute his comments for the purposes of both section 309 of the Clean Air Act and section 102(2)(C) of the National Environmental Policy Act.

(c) State and local review. Office of Management and Budget Circular No. A–95 (Revised) through its system of State and areawide clearinghouses provides a means for securnig the views of State and local environmental agencies, which can assist in the preparation and review of environmental impact statements. Current instructions for obtaining the views of such agencies are contained in the joint OMB–CEQ memorandum attached to these guidelines as Appendix IV. A current listing of clearinghouses is issued periodically by the Office of Management and Budget.

(d) *Public review.* The procedures established by these guidelines are designed to encourage public participation in the impact statement process at the earliest possible time. Agency procedures should make provision for facilitating the comment of public and private organizations and individuals by announcing the availability of draft environmental statements and by making copies available to organizations and individuals that request an opportunity to comment. Agencies should devise methods for publicizing the existence of draft statements, for example, by publication of notices in local newspapers or by maintaining a list of groups, including relevant conservation commissions, known to be interested in the agency's activities and directly notifying such groups of the existence of a draft statement, or sending them a copy, as soon as it has been prepared. A copy of the draft statement should in all cases be sent to any applicant whose project is the subject of the statement. Materials to be made available to the public shall be provided without charge to the extent practicable, or at a fee which is not more than the actual cost of reproducing copies required to be sent to other Federal agencies, including the Council.

(e) *Responsibilities of commenting entities.* (1) Agencies and members of the public submitting comments on proposed actions on the basis of draft environmental statements should endeavor to make their comments as specific, substantive, and factual as possible without undue attention to matters of form in the impact statement. Although the comments need not conform to any particular format, it would assist agencies reviewing comments if the comments were organized in a manner consistent with the structure of the draft statement. Emphasis should be placed on the assessment of the environmental impacts of the proposed action, and the acceptability of those impacts on the quality of the environment, particularly as contrasted with the impacts of reasonable alternatives to the action. Commenting entities may recommend modifications to the proposed action and/or new alternatives that will enhance environmental quality and avoid or minimize adverse environmental impacts.

(2) Commenting agencies should indicate whether any of their projects not identified in the draft statement are sufficiently advanced in planning and related environmentally to the proposed action so that a discussion of the environmental interrelationships should be included in the final statement (see § 1500.8(a)(1)). The Council is available to assist agencies in making such determinations.

(3) Agencies and members of the public should indicate in their comments the nature of any monitoring of the environmental effects of the proposed project that appears particularly appropriate. Such monitoring may be necessary during the construction, startup, or operation phases of the project. Agencies with special expertise with respect to the environmental impacts involved are encouraged to assist the sponsoring agency in the establishment and operation of appropriate environmental monitoring.

(f) Agencies seeking comment shall establish time limits of not less than forty-five (45) days for reply, after which it may be presumed, unless the agency or party consulted requests a specified extension of time, that the agency or party consulted has no comment to make. Agencies seeking comment should endeavor to comply with requests for extensions of time of up to fifteen (15) days. In determining an appropriate period for comment, agencies should consider the magnitude and complexity of the statement and the extent of citizen interest in the proposed action.

§ 1500.10 Preparation and circulation of final environmental statements.

(a) Agencies should make every effort to discover and discuss all major points of view on the environmental effects of the proposed action and its alternatives in the draft statement itself. However, where opposing professional views and responsible opinion have been overlooked in the draft statement and are brought to the agency's attention through the commenting process, the agency should review the environmental effects of the action in light of those views and should make a meaningful reference in the final statement to the existence of any responsible opposing view not adequately discussed in the draft statement, indicating the agency's response to the issues raised. All substantive comments received on the draft (or summaries thereof where response has been exceptionally voluminous) should be attached to the final statement, whether or not each such comment is thought to merit individual discussion by the agency in the text of the statement.

(b) Copies of final statements, with comments attached, shall be sent to all Federal, State, and local agencies and private organizations that made substantive comments on the draft statement and to individuals who requested a copy of the final statement, as well as any applicant whose project is the subject of the statement. Copies of final statements shall in all cases be sent to the Environmental Protection Agency to assist it in carrying out its responsibilities under section 309 of the Clean Air Act. Where the number of comments on a draft statement is such that distribution of the final statement to all commenting entities appears impracticable, the agency shall consult with the Council concerning alternative arrangements for distribution of the statement.

§ 1500.11 Transmittal of statements to the Council; minimum periods for review; requests by the Council.

(a) As soon as they have been prepared, ten (10) copies of draft environmental statements, five (5) copies of all comments made thereon (to be forwarded to the Council by the entity making comment at the time comment is forwarded to the responsible agency), and ten (10) copies of the final text of environmental statements (together with the substance of all comments received by the responsible agency from Federal, State, and local agencies and from private organizations and individuals) shall be supplied to the Council. This will serve to meet the statutory requirement to make environmental statements available to the President. At the same time that copies of draft and final statements are sent to the Council, copies should also be sent to relevant commenting entities as set forth in §§ 1500.9 and 1500.10(b) of these guidelines.

(b) To the maximum extent practicable no administrative action subject to section 102(2)(C) is to be taken sooner than ninety (90) days after a draft environmental statement has been circulated for comment, furnished to the Council and, except where advance public disclosure will result in significantly increased costs of procurement to the Government, made available to the public pursuant to these guidelines; neither should such administrative action[4] be taken sooner than thirty (30) days after the final text of an environmental statement (together with comments) has been made available to the Council, commenting agencies, and the public. In all cases, agencies should allot a sufficient review period for the final statement so as to comply with the statutory requirement that the "statement and the comments and views of appropriate Federal, State, and local agencies * * * accompany the proposal through the existing agency review processes." If the final text of an environmental statement is filed within ninety (90) days after a draft statement has been circulated for comment, furnished to the Council and made public pursuant to this section of these guidelines, the minimum thirty (30) day period and the ninety (90) day period may run concurrently to the extent that they overlap. An agency may at any time supplement or amend a draft or final environmental statement, particularly when substantial changes are made in the proposed action, or significant new information becomes available concerning its environmental aspects. In such cases the agency should consult with the Council with respect to the possible need for or desirability of recirculation of the statement for the appropriate period.

(c) The Council will publish weekly in the FEDERAL REGISTER lists of environmental statements received during the preceding week that are available for public comment. The date of publication of such lists shall be the date from which the minimum periods for review and advance availability of statements shall be calculated.

(d) The Council's publication of notice of the availability of statements is in addition to the agency's responsibility, as described in § 1500.9(d) of these guidelines, to insure the fullest practicable provision of timely public information concerning the existence and availability of environmental statements. The agency responsible for the environmental statement is also responsible for making the statement, the comments received, and any underlying documents available to the public pursuant to the provisions of the Freedom of Information Act (5 U.S.C., 552), without regard to the exclusion of intra- or interagency memoranda when such memoranda transmit comments of Federal agencies on the environmental impact of the proposed action pursuant to § 1500.9 of these guidelines. Agency procedures prepared pursuant to § 1500.3(a) of these guidelines shall implement these public information requirements and shall include arrangements for availability of environmental statements and comments at the head and appropriate regional offices of the responsible agency and at appropriate State and areawide clearinghouses unless the Governor of the State involved designates to the Council some other point for receipt of this information. Notice of such designation of an alternate point for receipt of this information will be included in the Office of Management and Budget listing of clearinghouses referred to in § 1500.9(c).

(e) Where emergency circumstances make it necessary to take an action with significant environmental impact without observing the provisions of these guidelines concerning minimum periods for agency review and advance availability of environmental statements, the Federal agency proposing to take the action should consult with the Council about alternative arrangements. Similarly where there are overriding considerations of expense to the Government or impaired program effectiveness, the responsible agency should consult with the Council concerning appropriate modifications of the minimum periods.

(f) In order to assist the Council in fulfilling its responsibilities under the Act and under Executive Order 11514, all agencies shall (as required by section 102(2)(H) of the Act and section 3(i) of Executive Order 11514) be responsive to requests by the Council for reports and other information dealing with issues arising in connection with the implementation of the Act. In particular, agencies shall be responsive to a request by the Council for the preparation and circulation of an environmental statement, unless the agency determines that such a statement is not required, in which case the agency shall prepare an environmental assessment and a publicly available record briefly setting forth the reasons for its determination. In no case, however, shall the Council's silence or failure to comment or request preparation, modification, or recirculation of an environmental statement or to take other action with respect to an environmental statement be construed as bearing in any way on the question of the legal requirement for or the adequacy of such statement under the Act.

§ 1500.12 Legislative actions.

(a) The Council and the Office of Management and Budget will cooperate in giving guidance as needed to assist agencies in identifying legislative items believed to have environmental significance. Agencies should prepare impact statements prior to submission of their legislative proposals to the Office of Management and Budget. In this regard, agencies should identify types of repetitive legislation requiring environmental impact statements (such as certain types of bills affecting transportation policy or annual construction authorizations).

(b) With respect to recommendations or reports on proposals for legislation to which section 102(2)(C) applies, the final text of the environmental statement and comments thereon should be available to the Congress and to the public for consideration in connection with the proposed legislation or report. In cases where the scheduling of congressional hearings on recommendations or reports on proposals for legislation which the Federal agency has forwarded to the Congress does not allow adequate time for the completion of a final text of an environmental statement (together with comments), a draft environmental statement may be furnished to the Congress and made available to the public pending transmittal of the comments as received and the final text.

§ 1500.13 Application of section 102(2)(C) procedure to existing projects and programs.

Agencies have an obligation to reassess ongoing projects and programs in order to avoid or minimize adverse environmental effects. The section 102(2)(C) procedure shall be applied to further major Federal actions having a significant effect on the environment even though they arise from projects or programs initiated prior to enactment of the Act on January 1, 1970. While the status of the work and degree of completion may be considered in determining whether to proceed with the project, it is essential that the environmental impacts of proceeding are reassessed pursuant to the Act's policies and procedures and, if the project or program is continued, that further incremental major actions be shaped so as to enhance and restore environmental quality as well as to avoid or minimize adverse environmental consequences. It is also important in further action that account be taken of environmental consequences not fully evaluated at the outset of the project or program.

§ 1500.14 Supplementary guidelines; evaluation of procedures.

(a) The Council after examining environmental statements and agency procedures with respect to such statements will issue such supplements to these guidelines as are necessary.

(b) Agencies will continue to assess their experience in the implementation of the section 102(2)(C) provisions of the Act and in conforming with these guidelines and report thereon to the Council by June 30, 1974. Such reports should include an identification of the problem areas and suggestions for revision or clarification of these guidelines to achieve effective coordination of views on environmental aspects (and alternatives, where appropriate) of proposed actions without imposing unproductive administrative procedures. Such reports shall also indicate what progress the agency has made in developing substantive criteria and guidance for making environmental assessments as required by § 1500.6(c) of this directive and by section 102(2)(B) of the Act.

Effective date. The revisions of these guidelines shall apply to all draft and final impact statements filed with the Council after January 28, 1973.

RUSSELL E. TRAIN,
Chairman.

APPENDIX I—SUMMARY TO ACCOMPANY DRAFT AND FINAL STATEMENTS

(Check one) ( ) Draft. ( ) Final Environmental Statement.

Name of responsible Federal agency (with name of operating division where appropriate). Name, address, and telephone number of individual at the agency who can be contacted for additional information about the proposed action or the statement.

1. Name of action (Check one) ( ) Administrative Action. ( ) Legislative Action.
2. Brief description of action and its purpose. Indicate what States (and counties) particularly affected, and what other proposed Federal actions in the area, if any, are discussed in the statement.
3. Summary of environmental impacts and adverse environmental effects.
4. Summary of major alternatives considered.
5. (For draft statements) List all Federal, State, and local agencies and other parties from which comments have been requested. (For final statements) List all Federal, State, and local agencies and other parties from which written comments have been received.
6. Date draft statement (and final environmental statement, if one has been issued) made available to the Council and the public.

APPENDIX II—AREAS OF ENVIRONMENTAL IMPACT AND FEDERAL AGENCIES AND FEDERAL STATE AGENCIES [1] WITH JURISDICTION BY LAW OR SPECIAL EXPERTISE TO COMMENT THEREON [2]

AIR

Air Quality

Department of Agriculture—
Forest Service (effects on vegetation)
Atomic Energy Commission (radioactive substances)
Department of Health, Education, and Welfare
Environmental Protection Agency
Department of the Interior—
Bureau of Mines (fossil and gaseous fuel combustion)
Bureau of Sport Fisheries and Wildlife (effect on wildlife)
Bureau of Outdoor Recreation (effects on recreation)
Bureau of Land Management (public lands)
Bureau of Indian Affairs (Indian lands)
National Aeronautics and Space Administration (remote sensing, aircraft emissions)
Department of Transportation—
Assistant Secretary for Systems Development and Technology (auto emissions)
Coast Guard (vessel emissions)
Federal Aviation Administration (aircraft emissions)

Weather Modification

Department of Agriculture—
Forest Service
Department of Commerce—
National Oceanic and Atmospheric Administration
Department of Defense—
Department of the Air Force
Department of the Interior
Bureau of Reclamation

WATER RESOURCES COUNCIL

WATER

Water Quality

Department of Agriculture—
Soil Conservation Service
Forest Service
Atomic Energy Commission (radioactive substances)
Department of the Interior—
Bureau of Reclamation
Bureau of Land Management (public lands)
Bureau of Indian Affairs (Indian lands)
Bureau of Sports Fisheries and Wildlife
Bureau of Outdoor Recreation
Geological Survey
Office of Saline Water
Environmental Protection Agency
Department of Health, Education, and Welfare
Department of Defense—
Army Corps of Engineers
Department of the Navy (ship pollution control)
National Aeronautics and Space Administration (remote sensing)
Department of Transportation—
Coast Guard (oil spills, ship sanitation)
Department of Commerce—
National Oceanic and Atmospheric Administration
Water Resources Council
River Basin Commissions (as geographically appropriate)

Marine Pollution, Commercial Fishery Conservation, and Shellfish Sanitation

Department of Commerce—
National Oceanic and Atmospheric Administration
Department of Defense—
Army Corps of Engineers
Office of the Oceanographer of the Navy
Department of Health, Education, and Welfare
Department of the Interior—
Bureau of Sport Fisheries and Wildlife
Bureau of Outdoor Recreation
Bureau of Land Management (outer continental shelf)
Geological Survey (outer continental shelf)
Department of Transportation—
Coast Guard
Environmental Protection Agency
National Aeronautics and Space Administration (remote sensing)
Water Resources Council
River Basin Commissions (as geographically appropriate)

Waterway Regulation and Stream Modification

Department of Agriculture—
Soil Conservation Service
Department of Defense—
Army Corps of Engineers
Department of the Interior—
Bureau of Reclamation
Bureau of Sport Fisheries and Wildlife
Bureau of Outdoor Recreation
Geological Survey
Department of Transportation—
Coast Guard
Environmental Protection Agency
National Aeronautics and Space Administration (remote sensing)
Water Resources Council
River Basin Commissions (as geographically appropriate)

FISH AND WILDLIFE

Department of Agriculture—
Forest Service
Soil Conservation Service
Department of Commerce—
National Oceanic and Atmospheric Administration (marine species)
Department of the Interior—
Bureau of Sport Fisheries and Wildlife
Bureau of Land Management
Bureau of Outdoor Recreation
Environmental Protection Agency

SOLID WASTE

Atomic Energy Commission (radioactive waste)
Department of Defense—
Army Corps of Engineers
Department of Health, Education, and Welfare
Department of the Interior—
Bureau of Mines (mineral waste, mine acid waste, municipal solid waste, recycling)
Bureau of Land Management (public lands)
Bureau of Indian Affairs (Indian lands)
Geological Survey (geologic and hydrologic effects)
Office of Saline Water (demineralization)
Department of Transportation—
Coast Guard (ship sanitation)
Environmental Protection Agency
River Basin Commissions (as geographically appropriate)
Water Resources Council

NOISE

Department of Commerce—
National Bureau of Standards
Department of Health, Education, and Welfare
Department of Housing and Urban Development (land use and building materials aspects)
Department of Labor—
Occupational Safety and Health Administration
Department of Transportation—
Assistant Secretary for Systems Development and Technology
Federal Aviation Administration, Office of Noise Abatement
Environmental Protection Agency
National Aeronautics and Space Administration

RADIATION

Atomic Energy Commission
Department of Commerce—
National Bureau of Standards
Department of Health, Education, and Welfare
Department of the Interior—
Bureau of Mines (uranium mines)
Mining Enforcement and Safety Administration (uranium mines)
Environmental Protection Agency

HAZARDOUS SUBSTANCES

Toxic Materials

Atomic Energy Commission (radioactive substances)
Department of Agriculture—
Agricultural Research Service
Consumer and Marketing Service
Department of Commerce—
National Oceanic and Atmospheric Administration
Department of Defense
Department of Health, Education, and Welfare
Environmental Protection Agency

---

[1] River Basin Commissions (Delaware, Great Lakes, Missouri, New England, Ohio, Pacific Northwest, Souris-Red-Rainy, Susquehanna, Upper Mississippi) and similar Federal-State agencies should be consulted on actions affecting the environment of their specific geographic jurisdictions.

[2] In all cases where a proposed action will have significant international environmental effects, the Department of State should be consulted, and should be sent a copy of any draft and final impact statement which covers such action.

Food Additives and Contamination of Foodstuffs

Department of Agriculture—
Consumer and Marketing Service (meat and poultry products)
Department of Health, Education, and Welfare
Environmental Protection Agency

Pesticides

Department of Agriculture—
Agricultural Research Service (biological controls, food and fiber production)
Consumer and Marketing Service
Forest Service
Department of Commerce—
National Oceanic and Atmospheric Administration
Department of Health, Education, and Welfare
Department of the Interior—
Bureau of Sport Fisheries and Wildlife (fish and wildlife effects)
Bureau of Land Management (public lands)
Bureau of Indian Affairs (Indian lands)
Bureau of Reclamation (irrigated lands)
Environmental Protection Agency

Transportation and Handling of Hazardous Materials

Atomic Energy Commission (radioactive substances)
Department of Commerce—
Maritime Administration
National Oceanic and Atmospheric Administration (effects on marine life and the coastal zone)
Department of Defense—
Armed Services Explosive Safety Board
Army Corps of Engineers (navigable waterways)
Department of Transportation—
Federal Highway Administration, Bureau of Motor Carrier Safety
Coast Guard
Federal Railroad Administration
Federal Aviation Administration
Assistant Secretary for Systems Development and Technology
Office of Hazardous Materials
Office of Pipeline Safety
Environmental Protection Agency

ENGERY SUPPLY AND NATURAL RESOURCES DEVELOPMENT

Electric Energy Development, Generation, and Transmission, and Use

Atomic Energy Commission (nuclear)
Department of Agriculture—
Rural Electrification Administration (rural areas)
Department of Defense—
Army Corps of Engineers (hydro)
Department of Health, Education, and Welfare (radiation effects)
Department of Housing and Urban Development (urban areas)
Department of the Interior—
Bureau of Indian Affairs (Indian lands)
Bureau of Land Management (public lands)
Bureau of Reclamation
Power Marketing Administrations
Geological Survey
Bureau of Sport Fisheries and Wildlife
Bureau of Outdoor Recreation
National Park Service
Environmental Protection Agency
Federal Power Commission (hydro, transmission, and supply)
River Basin Commissions (as geographically appropriate)
Tennessee Valley Authority
Water Resources Council

Petroleum Development, Extraction, Refining, Transport, and Use

Department of the Interior—
Office of Oil and Gas
Bureau of Mines
Geological Survey
Bureau of Land Management (public lands and outer continental shelf)
Bureau of Indian Affairs (Indian lands)
Bureau of Sport Fisheries and Wildlife (effects on fish and wildlife)
Bureau of Outdoor Recreation
National Park Service
Department of Transportation (Transport and Pipeline Safety)
Environmental Protection Agency
Interstate Commerce Commission

Natural Gas Development, Production, Transmission, and Use

Department of Housing and Urban Development (urban areas)
Department of the Interior—
Office of Oil and Gas
Geological Survey
Bureau of Mines
Bureau of Land Management (public lands)
Bureau of Indian Affairs (Indian lands)
Bureau of Sport Fisheries and Wildlife
Bureau of Outdoor Recreation
National Park Service
Department of Transportation (transport and safety)
Environmental Protection Agency
Federal Power Commission (production, transmission, and supply)
Interstate Commerce Commission

Coal and Minerals Development, Mining, Conversion, Processing, Transport, and Use

Appalachian Regional Commission
Department of Agriculture—
Forest Service
Department of Commerce
Department of the Interior—
Office of Coal Research
Mining Enforcement and Safety Administration
Bureau of Mines
Geological Survey
Bureau of Indian Affairs (Indian lands)
Bureau of Land Management (public lands)
Bureau of Sport Fisheries and Wildlife
Bureau of Outdoor Recreation
National Park Service
Department of Labor—
Occupational Safety and Health Administration
Department of Transportation
Environmental Protection Agency
Interstate Commerce Commission
Tennessee Valley Authority

Renewable Resource Development, Production, Management, Harvest, Transport, and Use

Department of Agriculture—
Forest Service
Soil Conservation Service
Department of Commerce
Department of Housing and Urban Development (building materials)
Department of the Interior—
Geological Survey
Bureau of Land Management (public lands)
Bureau of Indian Affairs (Indian lands)
Bureau of Sport Fisheries and Wildlife
Bureau of Outdoor Recreation
National Park Service
Department of Transportation
Environmental Protection Agency
Interstate Commerce Commission (freight rates)

Energy and Natural Resources Conservation

Department of Agriculture—
Forest Service
Soil Conservation Service
Department of Commerce—
National Bureau of Standards (energy efficiency)
Department of Housing and Urban Development—
Federal Housing Administration (housing standards)
Department of the Interior—
Office of Energy Conservation
Bureau of Mines
Bureau of Reclamation
Geological Survey
Power Marketing Administration
Department of Transportation
Environmental Protection Agency
Federal Power Commission
General Services Administration (design and operation of buildings)
Tennessee Valley Authority

LAND USE AND MANAGEMENT

Land Use Changes, Planning and Regulation of Land Development

Department of Agricuture—
Forest Service (forest lands)
Agricultural Research Service (agricultural lands)
Department of Housing and Urban Development
Department of the Interior—
Office of Land Use and Water Planning
Bureau of Land Management (public la
Bureau of Land Management (public lands)
Bureau of Indian Affairs (Indian lands)
Bureau of Sport Fisheries and Wildlife (wildlife refuges)
Bureau of Outdoor Recreation (recreation lands)
National Park Service (NPS units)
Department of Transportation
Environmental Protection Agency (pollution effects)
National Aeronautics and Space Administration (remote sensing)
River Basins Commissions (as geographically appropriate).

Public Land Management

Department of Agriculture—
Forest Service (forests)
Department of Defense
Department of the Interior—
Bureau of Land Management
Bureau of Indian Affairs (Indian lands)
Bureau of Sport Fisheries and Wildlife (wildlife refuges)
Bureau of Outdoor Recreation (recreation lands)
National Park Service (NPS units)
Federal Power Commission (project lands)
General Services Administration
National Aeronautics and Space Administration (remote sensing)
Tennessee Valley Authority (project lands)

PROTECTION OF ENVIRONMENTALLY CRITICAL AREAS—FLOODPLAINS, WETLANDS, BEACHES AND DUNES, UNSTABLE SOILS, STEEP SLOPES, AQUIFER RECHARGE AREAS, ETC.

Department of Agriculture—
Agricultural Stabilization and Conservation Service
Soil Conservation Service
Forest Service
Department of Commerce—
National Oceanic and Atmospheric Administration (coastal areas)
Department of Defense—
Army Corps of Engineers
Department of Housing and Urban Development (urban and floodplain areas)

Department of the Interior—
Office of Land Use and Water Planning
Bureau of Outdoor Recreation
Bureau of Reclamation
Bureau of Sport Fisheries and Wildlife
Bureau of Land Management
Geological Survey
Environmental Protection Agency (pollution effects)
National Aeronautics and Space Administration (remote sensing)
River Basins Commissions (as geographically appropriate)
Water Resources Council

LAND USE IN COASTAL AREAS

Department of Agriculture—
Forest Service
Soil Conservation Service (soil stability, hydrology)
Department of Commerce—
National Oceanic and Atmospheric Administration (impact on marine life and coastal zone management)
Department of Defense—
Army Corps of Engineers (beaches, dredge and fill permits, Refuse Act permits)
Department of Housing and Urban Development (urban areas)
Department of the Interior—
Office of Land Use and Water Planning
Bureau of Sport Fisheries and Wildlife
National Park Service
Geological Survey
Bureau of Outdoor Recreation
Bureau of Land Management (public lands)
Department of Transportation—
Coast Guard (bridges, navigation)
Environmental Protection Agency (pollution effects)
National Aeronautics and Space Administration (remote sensing)

REDEVELOPMENT AND CONSTRUCTION IN BUILT-UP AREAS

Department of Commerce—
Economic Development Administration (designated areas)
Department of Housing and Urban Development
Department of the Interior—
Office of Land Use and Water Planning
Department of Transportation
Environmental Protection Agency
General Services Administration
Office of Economic Opportunity

DENSITY AND CONGESTION MITIGATION

Department of Health, Education, and Welfare
Department of Housing and Urban Development
Department of the Interior—
Office of Land Use and Water Planning
Bureau of Outdoor Recreation
Department of Transportation
Environmental Protection Agency

NEIGHBORHOOD CHARACTER AND CONTINUITY

Department of Health, Education, and Welfare
Department of Housing and Urban Development
National Endowment for the Arts
Office of Economic Opportunity

IMPACTS ON LOW-INCOME POPULATIONS

Department of Commerce—
Economic Development Administration (designated areas)
Department of Health, Education, and Welfare
Department of Housing and Urban Development
Office of Economic Opportunity

HISTORIC, ARCHITECTURAL, AND ARCHEOLOGICAL PRESERVATION

Advisory Council on Historic Preservation
Department of Housing and Urban Development
Department of the Interior—
National Park Service
Bureau of Land Management (public lands)
Bureau of Indian Affairs (Indian lands)
General Services Administration
National Endowment for the Arts

SOIL AND PLANT CONSERVATION AND HYDROLOGY

Department of Agriculture—
Soil Conservation Service
Agricultural Service
Forest Service
Department of Commerce—
National Oceanic and Atmospheric Administration
Department of Defense—
Army Corps of Engineers (dredging, aquatic plants)
Department of Health, Education, and Welfare
Department of the Interior—
Bureau of Land Management
Bureau of Sport Fisheries and Wildlife
Geological Survey
Bureau of Reclamation
Environmental Protection Agency
National Aeronautics and Space Administration (remote sensing)
River Basin Commissions (as geographically appropriate)
Water Resources Council

OUTDOOR RECREATION

Department of Agriculture—
Forest Service
Soil Conservation Service
Department of Defense—
Army Corps of Engineers
Department of Housing and Urban Development (urban areas)
Department of the Interior—
Bureau of Land Management
National Park Service
Bureau of Outdoor Recreation
Bureau of Sport Fisheries and Wildlife
Bureau of Indian Affairs
Environmental Protection Agency
National Aeronautics and Space Administration (remote sensing)
River Basin Commissions (as geographically appropriate)
Water Resources Council

APPENDIX III—OFFICES WITHIN FEDERAL AGENCIES AND FEDERAL-STATE AGENCIES FOR INFORMATION REGARDING THE AGENCIES' NEPA ACTIVITIES AND FOR RECEIVING OTHER AGENCIES' IMPACT STATEMENTS FOR WHICH COMMENTS ARE REQUESTED

ADVISORY COUNCIL ON HISTORIC PRESERVATION

Office of Architectural and Environmental Preservation, Advisory Council on Historic Preservation, Suite 430, 1522 K Street, N.W., Washington, D.C. 20005 254–3974

DEPARTMENT OF AGRICULTURE [1]

Office of the Secretary, Attn: Coordinator Environmental Quality Activities, U.S. Department of Agriculture, Washington, D.C. 20250 447–3965

APPALACHIAN REGIONAL COMMISSION

Office of the Alternate Federal Co-Chairman, Appalachian Regional Commission, 1666 Connecticut Avenue, N.W., Washington, D.C. 20235 967–4103

DEPARTMENT OF THE ARMY (CORPS OF ENGINEERS)

Executive Director of Civil Works, Office of the Chief of Engineers, U.S. Army Corps of Engineers, Washington, D.C. 20314 693–7168

ATOMIC ENERGY COMMISSION

For nonregulatory matters: Office of Assistant General Manager for Biomedical and Environmental Research and Safety Programs, Atomic Energy Commission, Washington, D.C. 20545 973–3208
For regulatory matters: Office of the Assistant Director for Environmental Projects, Atomic Energy Commission, Washington, D.C. 20545 973–7531

DEPARTMENT OF COMMERCE

Office of the Deputy Assistant Secretary for Environmental Affairs, U.S. Department of Commerce, Washington, D.C. 20230 967–4335

DEPARTMENT OF DEFENSE

Office of the Assistant Secretary for Defense (Health and Environment), U.S. Department of Defense, Room 3E172, The Pentagon, Washington, D.C. 20301 697–2111

DELAWARE RIVER BASIN COMMISSION

Office of the Secretary, Delaware River Basin Commission, Post Office Box 360, Trenton, N.J. 08603 (609) 883–9500

ENVIRONMENTAL PROTECTION AGENCY [2]

Director, Office of Federal Activities, Environmental Protection Agency, 401 M Street, S.W., Washington, D.C. 20460 755–0777

[1] Requests for comments or information from individual units of the Department of Agriculture, e.g., Soil Conservation Service, Forest Service, etc. should be sent to the Office of the Secretary, Department of Agriculture, at the address given above.

[2] Contact the Office of Federal Activities for environmental statements concerning legislation, regulations, national program proposals or other major policy issues.

For all other EPA consultation, contact the Regional Administrator in whose area the proposed action (e.g., highway or water resource construction projects) will take place. The Regional Administrators will coordinate the EPA review. Addresses of the Regional Administrators, and the areas covered by their regions are as follows:

Regional Administrator, I, U.S. Environmental Protection Agency, Room 2303, John F. Kennedy Federal Bldg., Boston, Mass. 02203, (617) 223–7210	Connecticut, Maine, Massachusetts, New Hampshire, Rhode Island, Vermont
Regional Administrator, II, U.S. Environmental Protection Agency, Room 908, 26 Federal Plaza, New York, New York 10007, (212) 264–2525	New Jersey, New York, Puerto Rico, Virgin Islands

Regional Administrator, III, U.S. Environmental Protection Agency Curtis Bldg., 6th & Walnut Sts. Philadelphia, Pa. 19106 (215) 597–9801	Delaware, Maryland, Pennsylvania, Virginia, West Virginia, District of Columbia
Regional Administrator, IV, U.S. Environmental Protection Agency 1421 Peachtree Street N.E., Atlanta, Ga. 30309 (404) 526–5727	Alabama, Florida, Georgia, Kentucky Mississippi, North Carolina, South Carolina, Tennessee
Regional Administrator V, U.S. Environmental Protection Agency 1 N. Wacker Drive Chicago, Illinois 60606 (312) 353–5250	Illinois, Indiana, Michigan, Minnesota, Ohio, Wisconsin
Regional Administrator VI, U.S. Environmental Protection Agency 1600 Patterson Street Suite 1100 Dallas, Texas 75201 (214) 749–1962	Arkansas, Louisiana, New Mexico, Texas, Oklahoma
Regional Administrator VII, U.S. Environmental Protection Agency 1735 Baltimore Avenue Kansas City, Missouri 64108 (816) 374–5493	Iowa, Kansas, Missouri, Nebraska
Regional Administrator VIII, U.S. Environmental Protection Agency Suite 900, Lincoln Tower 1860 Lincoln Street Denver, Colorado 80203 (303) 837–3895	Colorado, Montana, North Dakota, South Dakota, Utah, Wyoming
Regional Administrator IX, U.S. Environmental Protection Agency 100 California Street San Francisco, California 94111 (415) 556–2320	Arizona, California, Hawaii, Nevada, American Samoa, Guam, Trust Territories of Pacific Islands, Wake Island
Regional Administrator X, U.S. Environmental Protection Agency 1200 Sixth Avenue Seattle, Washington 98101 (206) 442–1220	Alaska, Idaho, Oregon, Washington

FEDERAL POWER COMMISSION

Commission's Advisor on Environmental Quality, Federal Power Commission, 825 N. Capitol Street, N.E., Washington, D.C. 20426 386–6084

GENERAL SERVICES ADMINISTRATION

Office of Environmental Affairs, Office of the Deputy Administrator for Special Projects, General Services Administration, Washington, D.C. 20405 343–4161

GREAT LAKES BASIN COMMISSION

Office of the Chairman, Great Lakes Basin Commission, 3475 Plymouth Road, P.O. Box 999, Ann Arbor, Michigan 48105 (313) 769–7431

DEPARTMENT OF HEALTH, EDUCATION AND WELFARE [3]

Office of Environmental Affairs, Office of the Assistant Secretary for Administration and Management, Department of Health, Education and Welfare, Washington, D.C. 20202 963–4456

---

[3] Contact the Office of Environmental Affairs for information on HEW's environmental statements concerning legislation, regulations, national program proposals or other major policy issues, and for all requests for HEW comment on impact statements of other agencies.

For information with respect to HEW actions occurring within the jurisdiction of the Departments' Regional Directors, contact the appropriate Regional Environmental Officer:

Region I:
- Regional Environmental Officer
- U.S. Department of Health, Education and Welfare
- Room 2007B
- John F. Kennedy Center
- Boston, Massachusetts 02203 (617) 223–6837

Region II:
- Regional Environmental Officer
- U.S. Department of Health, Education and Welfare
- Federal Building
- 26 Federal Plaza
- New York, New York 10007 (212) 264–1308

Region III:
- Regional Environmental Officer
- U.S. Department of Health, Education and Welfare
- P.O. Box 13716
- Philadelphia, Pennsylvania 19101 (215) 597–6498

Region IV:
- Regional Environmental Officer
- U.S. Department of Health, Education and Welfare
- Room 404
- 50 Seventh Street, N.E.
- Atlanta, Georgia 30323 (404) 526–5817

Region V:
- Regional Environmental Officer
- U.S. Department of Health, Education and Welfare
- Room 712, New Post Office Building
- 433 West Van Buren Street
- Chicago, Illinois 60607 (312) 353–1644

DEPARTMENT OF HOUSING AND URBAN DEVELOPMENT [4]

Director, Office of Community and Environmental Standards, Department of Housing and Urban Development, Room 7206, Washington, D.C. 20410 755–5980

Region VI:
- Regional Environmental Officer
- U.S. Department of Health, Education and Welfare
- 1114 Commerce Street
- Dallas, Texas 75202 (214) 749–2236

Region VII:
- Regional Environmental Officer
- U.S. Department of Health, Education and Welfare
- 601 East 12th Street
- Kansas City, Missouri 64106 (816) 374–3584

Region VIII:
- Regional Environmental Officer
- U.S. Department of Health, Education and Welfare
- 9017 Federal Building
- 19th and Stout Streets
- Denver, Colorado 80202 (303) 837–4178

Region IX:
- Regional Environmental Officer
- U.S. Department of Health, Education and Welfare
- 50 Fulton Street
- San Francisco, California 94102 (415) 556–1970

Region X:
- Regional Environmental Officer
- U.S. Department of Health, Education and Welfare
- Arcade Plaza Building
- 1321 Second Street
- Seattle, Washington 98101 (206) 442–0490

[4] Contact the Director with regard to environmental impacts of legislation, policy statements, program regulations and procedures, and precedent-making project decisions. For all other HUD consultation, contact the HUD Regional Administrator in whose jurisdiction the project lies, as follows:

Regional Administrator I,
- Environmental Clearance Officer
- U.S. Department of Housing and Urban Development
- Room 405, John F. Kennedy Federal Building
- Boston, Mass. 02203 (617) 223–4066

Regional Administrator II,
- Environmental Clearance Officer
- U.S. Department of Housing and Urban Development
- 26 Federal Plaza
- New York, New York 10007 (212) 264–8068

Regional Administrator III,
- Environmental Clearance Officer
- U.S. Department of Housing and Urban Development
- Curtis Building, Sixth and Walnut Street
- Philadelphia, Pennsylvania 19106 (215) 597–2560

Regional Administrator IV,
- Environmental Clearance Officer
- U.S. Department of Housing and Urban Development
- Peachtree-Seventh Building
- Atlanta, Georgia 30323 (404) 526–5585

Regional Administrator V,
- Environmental Clearance Officer
- U.S. Department of Housing and Urban Development
- 360 North Michigan Avenue
- Chicago, Illinois 60601 (312) 353–5680

DEPARTMENT OF THE INTERIOR [5]

Director, Office of Environmental Project Review, Department of the Interior, Interior Building, Washington, D.C. 20240 343-3891

INTERSTATE COMMERCE COMMISSION

Office of Proceedings, Interstate Commerce, Commission, Washington, D.C. 20423
343-6167

DEPARTMENT OF LABOR

Assistant Secretary for Occupational Safety and Health, Department of Labor, Washington, D.C. 20210
961-3405

MISSOURI RIVER BASINS COMMISSION

Office of the Chairman, Missouri River Basins Commission, 10050 Regency Circle, Omaha, Nebraska 68114
(402) 397-5714

NATIONAL AERONAUTICS AND SPACE ADMINISTRATION

Office of the Comptroller, National Aeronautics and Space Administration, Washington, D.C. 20546
755-8440

NATIONAL CAPITAL PLANNING COMMISSION

Office of Environmental Affairs, Office of the Executive Director, National Capital Planning Commission, Washington, D.C. 20576
382-7200

NATIONAL ENDOWMENT FOR THE ARTS

Office of Architecture and Environmental Arts Program, National Endowment for the Arts, Washington, D.C. 20506
382-5765

NEW ENGLAND RIVER BASINS COMMISSION

Office of the Chairman, New England River Basins Commission, 55 Court Street, Boston, Mass. 02108
(617) 223-6244

Regional Administrator VI,
Environmental Clearance Officer
U.S. Department of Housing and Urban Development
Federal Office Building, 819 Taylor Street
Fort Worth, Texas 76102 (817) 334-2867

Regional Administrator VII,
Environmental Clearance Officer
U.S. Department of Housing and Urban Development
911 Walnut Street
Kansas City, Missouri 64106 (816) 374-2661

Regional Administrator VIII,
Environmental Clearance Officer
U.S. Department of Housing and Urban Development
Samsonite Building, 1051 South Broadway
Denver, Colorado 80209 (303) 837-4061

Regional Administrator IX,
Environmental Clearance Officer
U.S. Department of Housing and Urban Development
450 Golden Gate Avenue, Post Office Box 36003
San Francisco, California 94102 (415) 556-4752

Regional Administrator X,
Environmental Clearance Officer
U.S. Department of Housing and Urban Development
Room 226, Arcade Plaza Building
Seattle, Washington 98101 (206) 583-5415

[5] Requests for comments or information from individual units of the Department of the Interior should be sent to the Office of Environmental Project Review at the address given above.

OFFICE OF ECONOMIC OPPORTUNITY

Office of the Director, Office of Economic Opportunity, 1200 19th Street, N.W., Washington, D.C. 20506
254-6000

OHIO RIVER BASIN COMMISSION

Office of the Chairman, Ohio River Basin Commission, 36 East 4th Street, Suite 208-20, Cincinnati, Ohio 45202
(513) 684-3831

PACIFIC NORTHWEST RIVER BASINS COMMISSION

Office of the Chairman, Pacific Northwest River Basins Commission, 1 Columbia River, Vancouver, Washington 98660
(206) 695-3606

SOURIS-RED-RAINY RIVER BASINS COMMISSION

Office of the Chairman, Souris-Red-Rainy River Basins Commission, Suite 6, Professional Building, Holiday Mall, Moorhead, Minnesota 56560
(701) 237-5227

DEPARTMENT OF STATE

Office of the Special Assistant to the Secretary for Environmental Affairs, Department of State, Washington, D.C. 20520
632-7964

SUSQUEHANNA RIVER BASIN COMMISSION

Office of the Executive Director, Susquehanna River Basin Commission, 5012 Lenker Street, Mechanicsburg, Pa. 17055
(717) 737-0501

TENNESSEE VALLEY AUTHORITY

Office of the Director of Environmental Research and Development, Tennessee Valley Authority, 720 Edney Building, Chattanooga, Tennessee 37401 (615) 755-2002

DEPARTMENT OF TRANSPORTATION [6]

Director, Office of Environmental Quality, Office of the Assistant Secretary for Environment, Safety, and Consumer Affairs, Department of Transportation, Washington, D.C. 20590 426-4357

---

[6] Contact the Office of Environmental Quality, Department of Transportation, for information on DOT's environmental statements concerning legislation, regulations, national program proposals, or other major policy issues.

For information regarding the Department of Transportation's other environmental statements, contact the national office for the appropriate administration:

U.S. Coast Guard

Office of Marine Environment and Systems, U.S. Coast Guard, 400 7th Street, S.W., Washington, D.C. 20590, 426-2007

Federal Aviation Administration

Office of Environmental Quality, Federal Aviation Administration, 800 Independence Avenue, S.W., Washington, D.C. 20591, 426-8406

Federal Highway Administration

Office of Environmental Policy, Federal Highway Administration, 400 7th Street, S.W., Washington, D.C. 20590, 426-0351

Federal Railroad Administration

Office of Policy and Plans, Federal Railroad Administration, 400 7th Street, S.W., Washington, D.C. 20590, 426-1567

Urban Mass Transportation Administration

Office of Program Operations, Urban Mass Transportation Administration, 400 7th Street, S.W., Washington, D.C. 20590, 426-4020

For other administration's not listed above, contact the Office of Environmental Quality, Department of Transportation, at the address given above.

For comments on other agencies' environmental statements, contact the appropriate administration's regional office. If more than one administration within the Department of Transportation is to be requested to comment, contact the Secretarial Representative in the appropriate Regional Office for coordination of the Department's comments:

SECRETARIAL REPRESENTATIVE

Region I Secretarial Representative, U.S. Department of Transportation, Transportation Systems Center, 55 Broadway, Cambridge, Massachusetts 02142 (617) 494-2709

Region II Secretarial Representative, U.S. Department of Transportation, 26 Federal Plaza, Room 1811, New York, New York 10007 (212) 264-2672

Region III Secretarial Representative, U.S. Department of Transportation, Mall Building, Suite 1214, 325 Chestnut Street, Philadelphia, Pennsylvania 19106 (215) 597-0407

Region IV Secretarial Representative, U.S. Department of Transportation, Suite 515, 1720 Peachtree Rd., N.W. Atlanta, Georgia 30309 (404) 526-3738

Region V Secretarial Representative, U.S. Department of Transportation, 17th Floor, 300 S. Wacker Drive, Chicago, Illinois 60606 (312) 353-4000

Region V Secretarial Representative, U.S. Department of Transportation, 9-C-18 Federal Center, 1100 Commerce Street, Dallas, Texas 75202 (214) 749-1851

Region VII Secretarial Representative, U.S. Department of Transportation, 601 E. 12th Street, Room 634, Kansas City, Missouri 64106 (816) 374-2761

Region VIII Secretarial Representative, U.S. Department of Transportation, Prudential Plaza, Suite 1822, 1050 17th Street, Denver, Colorado 80225 (303) 837-3242

Region IX Secretarial Representative, U.S. Department of Transportation, 450 Golden Gate Avenue, Box 36133, San Francisco, California 94102 (415) 556-5961

Region X Secretarial Representative, U.S. Department of Transportation, 1321 Second Avenue, Room 507, Seattle, Washington 98101 (206) 442-0590

FEDERAL AVIATION ADMINISTRATION

New England Region, Office of the Regional Director, Federal Aviation Administration, 154 Middlesex Street, Burlington, Massachusetts 01803 (617) 272-2350

Eastern Region, Office of the Regional Director, Federal Aviation Administration, Federal Building, JFK International Airport, Jamaica, New York 11430 (212) 995-3333

Southern Region, Office of the Regional Director, Federal Aviation Administration, P.O. Box 20636, Atlanta, Georgia 30320 (404) 526-7222

Great Lakes Region, Office of the Regional Director, Federal Aviation Administration, 2300 East Devon, Des Plaines, Illinois 60018 (312) 694-4500

Southwest Region, Office of the Regional Director, Federal Aviation Administration, P.O. Box 1689, Fort Worth, Texas 76101 (817) 624-4911

Central Region, Office of the Regional Director, Federal Aviation Administration, 601 E. 12th Street, Kansas City, Missouri 64106 (816) 374-5626

Rocky Mountain Region, Office of the Regional Director, Federal Aviation Administration, Park Hill Station, P.O. Box 7213, Denver, Colorado 80207 (303) 837-3646

Western Region, Office of the Regional Director, Federal Aviation Administration, P.O. Box 92007, WorldWay Postal Center, Los Angeles, California 90009 (213) 536-6427

Northwest Region, Office of the Regional Director, Federal Aviation Administration, FAA Building, Boeing Field, Seattle, Washington 98108 (206) 767–2780

FEDERAL HIGHWAY ADMINISTRATION

Region 1, Regional Administrator, Federal Highway Administration, 4 Normanskill Boulevard, Delmar, New York 12054 (518) 472–6476

Region 3, Regional Administrator, Federal Highway Administration, Room 1621, George H. Fallon Federal Office Building, 31 Hopkins Plaza, Baltimore, Maryland 21201 (301) 962–2361

Region 4, Regional Administrator, Federal Highway Administration, Suite 200, 1720 Peachtree Road, N.W., Atlanta, Georgia 30309 (404) 526–5078

Region 5, Regional Administrator, Federal Highway Administration, Dixie Highway, Homewood, Illinois 60430 (312) 799–6300

Region 6, Regional Administrator, Federal Highway Administration, 819 Taylor Street, Fort Worth, Texas 76102 (817) 334–3232

Region 7, Regional Administrator, Federal Highway Administration, P.O. Box 7186, Country Club Station, Kansas City, Missouri 64113 (816) 361–7563

Region 8, Regional Administrator, Federal Highway Administration, Room 242, Building 40, Denver Federal Center, Denver, Colorado 80225

Region 9, Regional Administrator, Federal Highway Administration, 450 Golden Gate Avenue, Box 36096, San Francisco, California 94102 (415) 556–3895

Region 10, Regional Administrator, Federal Highway Administration, Room 412, Mohawk Building, 222 S.W. Morrison Street, Portland, Oregon 97204 (503) 221–2065

URBAN MASS TRANSPORTATION ADMINISTRATION

Region I, Office of the UMTA Representative, Urban Mass Transportation Administration, Transportation Systems Center, Technology Building, Room 277, 55 Broadway, Boston, Massachusetts 02142 (617) 494–2055

Region II, Office of the UMTA Representative, Urban Mass Transportation Administration, 26 Federal Plaza, Suite 1809, New York, New York 10007 (212) 264–8162

Region III, Office of the UMTA Representative, Urban Mass Transportation Administration, Mall Building, Suite 1214, 325 Chestnut Street, Philadelphia, Pennsylvania 19106 (215) 597–0407

Region IV, Office of UMTA Representative, Urban Mass Transportation Administration, 1720 Peachtree Road, Northwest, Suite 501, Atlanta, Georgia 30309 (404) 526–3948

Region V, Office of the UMTA Representative, Urban Mass Transportation Administration, 300 South Wacker Drive, Suite 700, Chicago, Illinois 60606 (312) 353–6005

Region VI, Office of the UMTA Repreventative, Urban Mass Transportation Administration, Federal Center, Suite 9E24, 1100 Commerce Street, Dallas, Texas 75202 (214) 749–7322

Region VII, Office of the UMTA Representative, Urban Mass Transportation Administration, c/o FAA Management Systems Division, Room 1564D, 601 East 12th Street, Kansas City, Missouri 64106 (816) 374–5567

Region VIII, Office of the UMTA Representative, Urban Mass Transportation Administration, Prudential Plaza, Suite 1822, 1050 17th Street, Denver, Colorado 80202 (303) 837–3242

Region IX, Office of the UMTA Representative, Urban Mass Transportation Administration, 450 Golden Gate Avenue, Box 36125, San Francisco, California 94102 (415) 556–2884

Region X, Office of the UMTA Representative, Urban Mass Transportation Administration, 1321 Second Avenue, Suite 5079, Seattle, Washington (206) 442–0590

DEPARTMENT OF THE TREASURY

Office of Assistant Secretary for Administration, Department of the Treasury, Washington, D.C. 20220 964–5391

UPPER MISSISSIPPI RIVER BASIN COMMISSION

Office of the Chairman, Upper Mississippi River Basin Commission, Federal Office Building, Fort Snelling, Twin Cities, Minnesota 55111 (612) 725–4690

WATER RESOURCES COUNCIL

Office of the Associate Director, Water Resources Council, 2120 L Street, N.W., Suite 800, Washington, D.C. 20037 254–6442

APPENDIX IV—STATE AND LOCAL AGENCY REVIEW OF IMPACT STATEMENTS

1. OMB Circular No. A–95 through its system of clearinghouses provides a means for securing the views of State and local environmental agencies, which can assist in the preparation of impact statements. Under A–95, review of the proposed project in the case of federally assisted projects (Part I of A–95) generally takes place prior to the preparation of the impact statement. Therefore, comments on the environmental effects of the proposed project that are secured during this stage of the A–95 process represent inputs to the environmental impact statement.

2. In the case of direct Federal development (Part II of A–95), Federal agencies are required to consult with clearinghouses at the earliest practicable time in the planning of the project or activity. Where such consultation occurs prior to completion of the draft impact statement, comments relating to the environmental effects of the proposed action would also represent inputs to the environmental impact statement.

3. In either case, whatever comments are made on environmental effects of proposed Federal or federally assisted projects by clearinghouses, or by State and local environmental agencies through clearinghouses, in the course of the A–95 review should be attached to the draft impact statement when it is circulated for review. Copies of the statement should be sent to the agencies making such comments. Whether those agencies then elect to comment again on the basis of the draft impact statement is a matter to be left to the discretion of the commenting agency depending on its resources, the significance of the project, and the extent to which its earlier comments were considered in preparing the draft statement.

4. The clearinghouses may also be used, by mutual agreement, for securing reviews of the draft environmental impact statement. However, the Federal agency may wish to deal directly with appropriate State or local agencies in the review of impact statements because the clearinghouses may be unwilling or unable to handle this phase of the process. In some cases, the Governor may have designated a specific agency, other than the clearinghouse, for securing reviews of impact statements. In any case, the clearinghouses should be sent copies of the impact statement.

5. To aid clearinghouses in coordinating State and local comments, draft statements should include copies of State and local agency comments made earlier under the A–95 process and should indicate on the summary sheet those other agencies from which comments have been requested, as specified in Appendix I of the CEQ Guidelines.

[FR Doc.73–15783 Filed 7–31–73;8:45 am]

BOOKS PUBLISHED BY EDRC

ENERGY

Lee, Dr. Kaiman / Masloff, Jacqueline
Kaiman's Encyclopedia Of Energy Topics
Copyright, January 1979, 2 Volumes, 800 p.
ISBN 0-915250-31-4

Lee, Dr. Kaiman / Masloff, Jacqueline
Energy-Oriented Computer Programs For The Design & Monitoring Of Buildings
Copyright, January 1979, 2 Volumes, 400 p.
ISBN 0-915250-29-2

Lee, Dr. Kaiman / Silverstein, Michael
The Buyer's Book Of Solar Water Heaters
Copyright, January 1979, 4th Edition, 208 p.
ISBN 0-915250-30-6

Lee, Dr. Kaiman
Encyclopedia Of Energy-Efficient Building Design: 391 Practical Case Studies
Copyright, April 1977, 2 Volumes, 1053 p.
LC 77-150686 TJ163.5.B84L43 ISBN 0-915250-18-7

Lee, Dr. Kaiman / Rehr, Stuart L.
Energy Conservation And Building Codes: The Legislative And Planning Processes
Copyright, April 1977, 109 p.
LC 77-362518 KF5701.L43 ISBN 0-915250-23-3

Lee, Dr. Kaiman
Bibliography Of Energy Conservation In Architecture: Keyword Searched
Copyright, April 1977, 428 p.
LC 77-362422 NA2542.3 Z5943.E5L4
ISBN 0-915250 19-5

Lee, Dr. Kaiman
Energy Film List
Copyright, April 1977, Looseleaf, 7 p.
ISBN 0-915250-25-X

COMPUTER APPLICATIONS IN ARCHITECTURE

Lee, Dr. Kaiman
Interactive Computer Graphics In Architecture
Copyright, May 1976, 100 p.
LC 76-366950 NA2728.L436 729'.01'83
ISBN 0-915250-21-7

Lee, Dr. Kaiman
Computer Aided Space Planning
Copyright, May 1976, 100 p.
LC 76-366704 NA2728.L432 729'.01'83
ISBN 0-915250-20-9

Lee, Dr. Kaiman
Interactive Computer Aided Architectural Design:
Four Applications
Published by Tamkeng College, Taiwan, 1976, 117 p.
ISBN 0-915250-26-8

Lee, Dr. Kaiman
Performance Specification Of Computer Aided
Environmental Design
Copyright, February 1975, 2 Volumes, 554 p.
LC 75-309149 NA2728.L437 729'.028'54
ISBN 0-915250-15-2

Lee, Dr. Kaiman
State Of The Art Of Computer Aided Environmental
Design
Copyright, February 1975, 309 p.
LC 76-358975 NA2728.L437 720'.28'54
ISBN 0-915250-14-4

Lee, Dr. Kaiman
Computer Programs In Environmental Design
Copyright, January 1974, 5 Volumes, 1308 p.
LC 74-169212 NA1996.5.L43 729'.028'54
ISBN 0-915250-05-5

Lee, Kaiman
A Step Towards An Integrated Computer Aided Design
System For The Architect/Planner
Copyright, January 1974, 225 p.
LC 74-182899 NA2728.L44 729'.028'54
ISBN 0-915250-07-1

Lee, Kaiman
Computer-Aided Building Code Checking - A Demonstration
Copyright, January 1974, 34 p.
LC 75-308127 RA967.L43 343.07'8
ISBN 0-915250-08-X

Lee, Kaiman
Computer-Generated Perspective Drawings
Copyright, January 1974, 50 p.
LC 75-305512 NA2728.L435 720'.28
ISBN 0-915250-09-8

Lee, Kaiman
Evaluation, Syntheses And Development Of An Interactive
Approach To Space Allocation
Copyright, January 1974, 50 p.
LC 74-187884 NA2765.L43 729'.028'54
ISBN 0-915250-10-1

Lee, Kaiman
Computer Aided Architectural Design: 16 Ark-2 Articles
Published By The Center For Environmental Research,
Copyright, December 1973, Approximately 150 p.
LC 74-160961 NA2728.L43 729'.028'54
ISBN 0-915250-04-7

Lee, Kaiman
Bibliography Of The Computer In Environmental Design
Published By The Center For Environmental Research,
Copyright, August 1973, 2nd Edition, 3 Volumes,
Approximately 650 p.
LC 73-158197 Z5863.E57.L43 016.711'028'54
ISBN 0-915250-03-9

Lee, Kaiman
The Computer As An Architectural Design Tool: An
Exploration Into Certain Multi-Story Building Plan
Layouts
Copyright, August 1969, 190 p.
ISBN 0-915250-00-4

Lee, Kaiman
Computerized Multi-Story Building Layout - COMSBUL
Computer Program
A Source Deck Of Cards And A Documentation Manual,
August 1969.

Big Production, Inc.
The Computer In Architecture
16 mm. Color And Sound Film, 15 min., Copyright 1970.

Lee, Kaiman
Ark-2: Computer Aided Architectural Design System
Slides and Text.
Single Screen Set of 80
Twin Screen Set Of 160

ENVIRONMENT

Lee, Dr. Kaiman / Koumjian, Lauren
Environmental Court Cases Related To Buildings
Copyright, January 1979, 113 p.
ISBN 0-915250-28-4

Lee, Kaiman / Koumjian, Lauren
Environmental Impact Statement: A Reference Manual For The Architect/Planner
Copyright, July 1978, 2nd Edition, 142 p.
ISBN 0-915250-27-6

Lee, Dr. Kaiman
Federal Environmental Impact Statements Related To Buildings
Copyright, June 1975, 120 p.
LC 75-323048 Z5863.E56L4 TD194.5
ISBN 0-015250-16-0

Lee, Dr. Kaiman / Moberg, John
Environmental Design Evaluation: A Matrix Method
Copyright, October 1975, 111 p.
LC 75-330795 TD194.5L42 363.6
ISBN 0-915250-17-9

Lee, Dr. Kaiman
Environmental Impact Information Matrix
A Chart Of 400 Existing Environmental Characteristics And 200 Proposed Actions, Size Of 3 Feet By 6 Feet.

Lee, Kaiman
Air Pollution: Its Effect On The Urban Man And His Adaptive Strategies
Copyright, January 1974, 52 p.
LC 74-182905 TD883.1.L4 641.7'1
ISBN 0-915250-13-6

OTHERS

Lee, Dr. Kaiman
Comparison Of Doctoral Programs In Architecture (USA)
Copyright, 1976, a chart of 24" x 36".
ISBN 0-915250-22-5

Lee, Kaiman
Integrated Municipal Information System
Copyright, January 1974, 52 p.
LC 74-184835 JS344.E4L43 352'.088'02854
ISBN 0-915250-12-8

Stewart, Clifford / Lee, Kaiman / Finne, Hans H.
Comprogram: Hospital Space Allocation
Published By The Center For Environmental Research,
Copyright, March 1972, 2 Volumes, Approximately 450 p.
LC 73-158290 RA967.S74 362.1'1
ISBN 0-915250-02-0

Lee, Kaiman
Evaluation Of Computer Graphic Terminals
Copyright, January 1974, 92 p.
LC 74-184824 TK7887.5.L4 621.3819'532
ISBN 0-915250-11-X

All are hard cover books except where indicated otherwise.

Publications available from:

Environmental Design + Research Center
142 Lowell Ave., Newtonville, MA 02160